DRAINAGE SYSTEMS

BASIC HANDBOOK

Kruger Brentt
Publishers

DRAINAGE SYSTEMS
BASIC HANDBOOK

Jonathan Chester
Adrian Pegg

Kruger Brentt
Publishers
2025

Kruger Brentt Publishers UK. LTD.
Company Number 9728962
Regd. Office: 68 St Margarets Road, Edgware, Middlesex HA8 9UU

ISBN: 978-1-78715-280-9

For information on all our publications visit our website at http://krugerbrentt.com/

PREFACE

Drainage systems are fundamental components of urban and rural infrastructure, serving to manage surface water runoff, prevent flooding, and maintain environmental quality. From stormwater drainage in urban areas to agricultural drainage in rural landscapes, drainage systems play a crucial role in safeguarding communities, protecting property, and sustaining ecosystems. Understanding the principles, design considerations, and maintenance practices of drainage systems is essential for engineers, planners, policymakers, and land managers involved in water resource management and infrastructure development.

This handbook, titled "Drainage Systems: Basic Handbook," is a comprehensive guide to the principles, practices, and applications of drainage systems. Drawing upon the expertise of engineers, hydrologists, and environmental scientists, this handbook offers practical guidance, design standards, and case studies for understanding and implementing effective drainage solutions.

From the fundamentals of hydrology and hydraulics to the design and construction of drainage networks, "Drainage Systems: Basic Handbook" covers a wide range of topics essential for professionals involved in drainage system planning, design, and management. Through clear explanations, diagrams, and examples, readers will gain a deep understanding of the principles and practices used to design and operate drainage systems in various contexts.

Moreover, this handbook explores the integration of drainage systems with other elements of water management, such as green infrastructure, sustainable urban drainage systems (SUDS), and ecosystem-based approaches. By examining case studies and best practices from around the world, we highlight innovative approaches to drainage system design that promote resilience, sustainability, and environmental stewardship.

As we confront the challenges of climate change, urbanization, and population growth, the importance of effective drainage systems has never been greater. It is our hope that "Drainage Systems: Basic Handbook" will serve as a valuable resource for professionals and students in the field of water resources engineering, environmental management, and urban planning. Whether you are designing drainage systems for urban development, managing agricultural drainage in rural landscapes, or implementing stormwater management projects, this handbook invites you to explore the principles and practices of drainage systems and contribute to the sustainable management of water resources for future generations.

Jonathan Chester
Adrian Pegg

CONTENTS

01 PALEODRAINAGE SYSTEMS

1. INTRODUCTION: AN OVERVIEW

River systems can be described as a series of water flow activities that occur at various scales in terms of space and time. These processes are structured within a given geographic location and are influenced by the climate, resulting in a state of dynamic balance. At a more specific and intricate level, there is a presence of turbulent water movement that is confined to a particular area within a channel. This movement can be accurately represented using equations that account for the local flow rate. At a broader level, it is important to evaluate other factors and processes that may affect sediment flow, such as the dynamics of sediment movement, the slope of the region, and variations over longer periods of time (seasonal and multi-annual). Open channel hydraulic methods can be used to model them. From a broader perspective, one can examine the large-scale basin and geological factors, as well as the regional and global climatic conditions, that influence long-term dynamics. In this chapter, the concept of paleodrainage is explained. It suggests that significant changes in river features are connected to larger environmental processes that can fundamentally transform the basin system. While local variations in the drainage system, such as seasonal flooding or small-scale lateral migration, may lead to different interpretations of terms like "paleovalley" and "paleochannel," it is generally accepted that only dynamic processes resulting in substantial geographical changes in drainage systems fall within this range of definitions. These changes are typically attributed to the most common causes:

- Downstream drowning/exposure (base level variation)
- Drying (climate change)
- Shifting (tectonic and morphosedimentary topographic induced processes)

These processes do not always happen as separate incidents, but rather as a series of environmental changes that involve fluctuations in sediment supply, topographic and geological conditions, local and regional climate, vegetation coverage, and so on. In this regional context, most studies on paleodrainage rely on two types of analysis: the geomorphological approach, which examines the response of landforms to drainage processes (as described in works by Al Sulaimi et al. 1997, Simpson 2004, and Dollar 2004), and the stratigraphic approach, which examines the sedimentary records of fluvial sequences within conceptual frameworks such as Quaternary Time Stratigraphy (Wheeler 1958) and Fluvial Sequence Stratigraphy (Wilgus et al. 1988; Strong and Paola 2008).

In addition, the analysis of ancient drainage patterns is typically associated with the examination of large-scale environmental conditions, particularly those related to Quaternary glacial cycles and tectonic processes. However, it is also used to provide context for historical trends and to predict future conditions in a changing climate, as well as human intervention in land cover scenarios. Understanding the dynamic evolution of natural drainage systems can offer valuable insights into the development of sustainable methodologies that integrate innovative water resource management and land use planning, which is the primary focus of this book.

2. PALEODRAINAGE AND ENVIRONMENTAL CHANGES: THE CAUSES OF RIVER FLUCTUATIONS ?

2.1 Tectonic control and paleodrainage

The literature extensively discusses the influence of tectonic activity on changes in drainage patterns, typically focusing on large geographic scales such as regions and continents. The impact of drainage on both the pattern and texture of the drainage system, as well as the quantity and properties of sediment transferred to downstream depositional basins and their delivery locations, is highly significant and might be regarded as the most striking consequence. Uplift, subsidence, and faulting can all change the general layout of drainage networks by affecting factors such as slope, valley floor, and channel gradient. The nature and extent of these changes will be determined by the degree of deformation and the ability of the channels to adapt to the modified slope (Shumm, 1977; 1985). Therefore, in order to comprehend the influence of tectonic evolution on the formation, evolution, and preservation of drainage systems, it is essential to investigate the interaction between externally induced (allogenic) and internally generated (autogenic) forces.

Burbank and Anderson (2001) provide examples of how tectonic events, particularly those related to orogenic belts, influence the development of drainage systems. An exemplary illustration of the correlation between tectonic processes and alterations in drainage patterns on a significant magnitude can be observed in the context of the rivers in Patagonia, located at the southern extremity of South America, as detailed in the works of Wilson (1991) and Tankard et al. (1995). The process of rifting that caused the separation of South America and Africa initially took place in the southern region of the continent during the Middle Jurassic period. The formation of wide uplifts, rifting, and aulacogens had a crucial role in determining the drainage patterns. The ancient course of the paleo Colorado and

Negro rivers, for instance, was linked to the Mesozoic paleosurfaces originating from the east and north of the Neuquen basin, flowing towards the Pacific Ocean. The uplift of the Andes arc during the Upper Cretaceous period led the slope to change direction towards the southeast. As a result, the rivers began to flow towards the Atlantic Ocean, passing along the northern side of the Somuncura uplift (Potter, 1997).

The Andean uplift also has controlled the paleo-geographic settings of all drainage in the central and northern cordillera region modifying the position of rivers such as the Orinoco and Magdelena at northwestern part of South America. As described in Horn *et al.* 1995 and Diaz de Gamero, 1996, during the Middle Eocen, the Orinoco River had been flowing south- north, draining the Central Cordillera of Colombia and the Guayana Highlands, with its mouth associated with the Lake Maracaibo and Caribbean Sea. After the Late Miocene period, the Andes attained their present configuration following the Eastern Cordillera uplift and the southwestern end of the Merida Andes deformation shifted the course of the river, deflecting it to a west-east direction. The eastern and central Cordillera uplift also controlled the paleo-Amazon River, which was connected to the Paleo-Orinoco until the mid Miocene period. The final breakthrough of the Amazon River towards its modern course occurred with the final uplift of the central Andean cordillera, related to the rise of the Purus Arch (Gregory-Wodzicky, 2000; Figueredo *et al.*, 2009) (Figure 1).

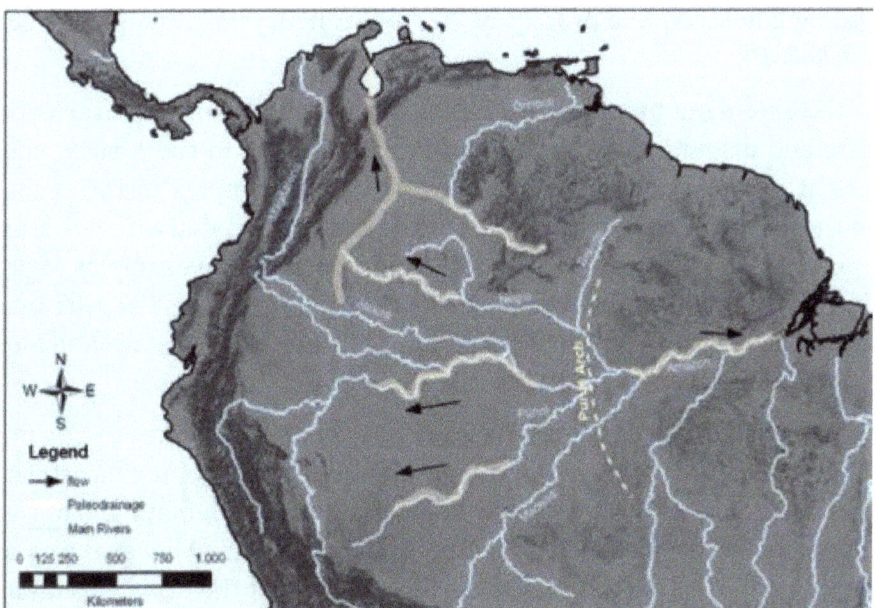

Fig. 1. Drainage distribution of Northern and Central areas of South America indicating the Amazon/Andes paleodrainage system (and flow direction) during Oligocene (modified after Potter, 1997 and Horn *et al.* 2010).

Tectonic processes have also controlled drainage evolution and basin infill trends in all active zones in the world. In the Himalayas, the Orogen began forming roughly 50 million years ago when India collided with the Asian continent producing complex relationships between regional metamorphism, anatexis, thrust faulting and normal faulting (Hodges, 2000). The establishment of the drainage systems and the evolutionary history of the fluvial network were controlled by this complex scenario characterized by both local and regional structures with multiple phases of deformation and shaped by active erosional dissection. Studies such as those done by Brookfield (1998), Zeitler *et al*, (2001), Najman *et al.* (2003), Clift and Blusztajn (2005), among others, affirmed that the transpressional tectonics across the Himalayan arc have produced large variations in fluvial morphodynamic processes in rivers such as the Indus, Kunar, Marsyangdi and Ganges. Most of these changes as described in Clift *et al.* (2006) can be related to capture/reversal processes.

Furthermore, the climatic and erosional processes related to such uplifts, which have determined the resultant landscape structure, are as important as the direct upward topographic responses to tectonic forces. Finnegan *et al.* (2008) analyzing the balance between uplift of the Namche Barwa massif in Tibet related to the Yarlung Tsangpo–Brahmaputra River system, concluded that the regional denudation driven by extreme efficient river incision, removes a large quantity of sediments yielding to a local relief producing erosional and thermo-mechanical coupling in an active orogen, thrusting Namche Barwa-Gyala Peri above the surrounding landscape in a process of "tectonic aneurysm" (Koons *et al.* 2002 and Simpson, 2004).

The tectonic dynamics can also affect large-scale drainage patterns through volcanism and distention processes in divergent plates. In the African continent there are at least two examples of rivers indicating complex histories involving shifts in the nature of their catchments and courses, associated with volcanism and rifting process. The Nile may be one of the most remarkable examples of drainage evolution in response to non-orogenic tectonic deformation. The Nile evolution model described by Sahin (1985), Foucault and Stanley (1989); Issawi and McCauley (1992) and Goudie (2004), details five phases since the late Eocene.

The first phase (called the "Gilf system" - Eocene) was marked by the presence of a set of parallel north flowing drainages occupying most of northeastern Africa and the present Nile Basin position fed by streams originating from eastern highlands (Red Sea area). The second Stage (Qena system - Miocene) was marked by intense tectonic activity followed by the reversion of the drainages taking up a southward position as well as establishing an immense river flowing south towards Aswan and the Sudan. The third stage (the Nile system – Late Miocene) was marked by a dramatic drop of the base level of the Mediterranean Sea (associated with the closure of the Straits of Gibraltar and the Messien Event – Gautier *et al.*, 1994) producing intense erosional activity forming gorges and canyons (designated as Eonile). As a result of such shifts, the Eonile took a northerly course. The fourth phase (Pliocene flooding), in the early Pliocene, the sea level rose at least 125 m so that an estuary or ria extended more than 900 km inland, reaching Aswan. After the Pleistocene, the Nile has been submitted to constant changes due to climate and hydrological regimes (detailed in Foucault and Stanley, 1985; Coutelier and Stanley, 1997).

The other example of a paleodrainage system controlled by distensive tectonics is the Zambezi River. The longitudinal profile of the Zambezi River forms two concave-upward sections, with their boundary at Victoria Falls, and there are studies suggesting that these sections were disconnected fluvial systems only joined together after the late Quaternary (Goudie, 2004; Nugent, 1990). The paleo-Zambezi was connected to both the Limpopo and Luangwa River system and took its present configuration, probably after the Pliocene with the uplift of the Makagdikgadi Basin along the "Kalahari Zimbabwe axis". The new position of the Middle Zambezi permitted the capturing of east and northeast tributaries such as the Kafue and Luangwa (Thomas and Shaw, 1991) (Figure 2).

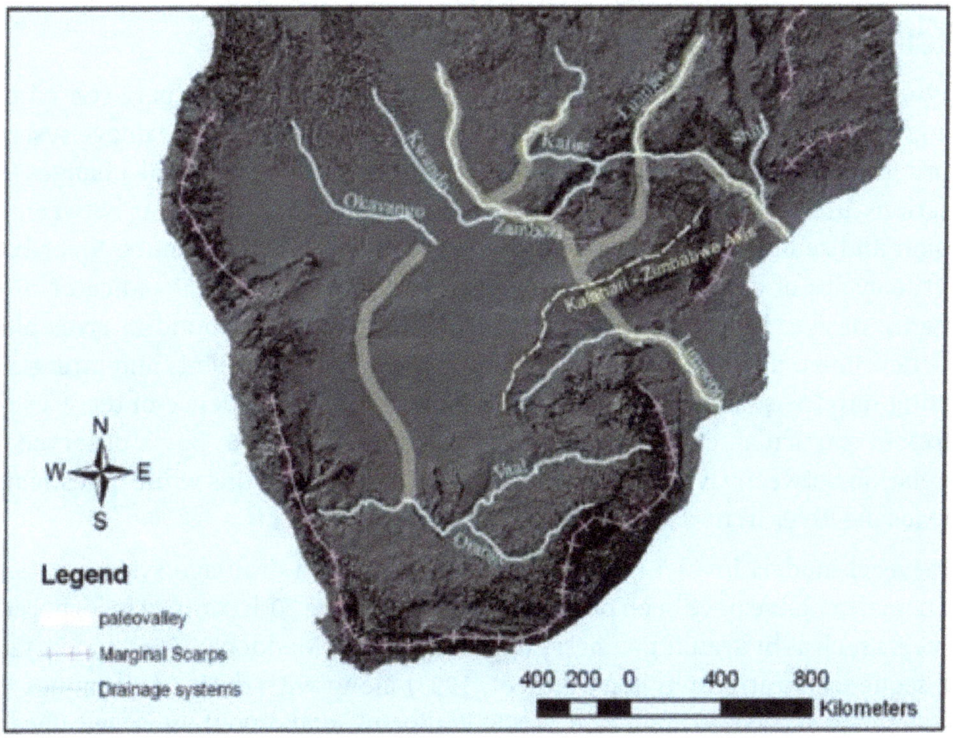

Fig. 2. Drainage distribution of Southern Africa with the paleodrainage prior to the union of the middle and upper Zambezi in the early Pleistocene (modified after Thomas and Shaw, 1991 and Goudie, 2005).

The evolutionary history of the Congo River is also linked to the tectonic processes related to south-central Africa vertical crust movements. Stankiewicz & Dewit (2006) suggested that the Congo drained into the Indian Ocean until the uplift of East African Highlands in the Oligocene or Eocene (30– 40 Ma). The Congo then became a landlocked basin, until it was captured during the Miocene by a river system causing it to drain into the Atlantic.

Examples of drainage modifications in catchments affected by the formation and erosion of volcanoes can be found in several scales, from regional drainage system displacements associated with major volcanic structures (such as *Serra Geral*/Etendeka and Deccan) to

localized events displacing and rearranging local streams. Oilier (1995), describes the relationship between directions of rivers in southeastern Australia (e.g. Hunter, Lachlan and Murrumbidgee) which reversed courses due to an Eocene/ early Oligocene Volcanism effect. In a more specific context, Branca and Ferrara, in 2001, created a map of the historic valley formed by the Alcantara and S. Paolo Rivers, located beneath the northern side of Mt Etna. The paleo-valley had multiple episodes of lava flow and deposition of volcanic sediment, which occurred between 170,000 years B.P. and 20,000 years B.P.

2.2 Climatic and hydrologic changes

Early empirical studies aimed to establish a direct correlation between climate changes and the settlement of morphosedimentary processes in drainage systems, in order to understand the relationship between environmental changes and variations in fluvial patterns. Penk (1927) suggested a connection between the erosion and sedimentation patterns of river channels in the Danube River basin and the cycles of glacial climate. This later became a significant indicator of the presence of glacial cycles. As per the authors, the terraces found in river plains were developed due to a cycle of glacial (building up of sediment) and interglacial (cutting into the sediment) periods. This cycle created a sequence of terraces that represent particular environmental circumstances. In 1896, Davis observed the correlation between river displacement and climate variations while investigating the Moselle River in northern France.

Several models involving the relationship between drainage system patterns and climate change have been proposed throughout the 20th century incorporating concepts such as hydraulic geometry (e.g. Leopold and Maddock, 1956), fluvial facies and sequence stratigraphy (e.g. Wescot, 1993) along with dating techniques (e.g. Sanderson *et al.*, 2003; Zhang *et al.*, 2003, Wallbrink *et al.*, 2002). In recent decades, multidisciplinary studies have proposed a more holistic approach to integrate climate change and paleodrainage pattern response affected by multiple drivers of varying spatial and temporal dominance and complexity (Knox, 1993, Ely *et al.*

1993; Gibbard and Lewin, 2002; Andres *et al.* 2002; Goodbred, 2003, Vandenberghe, 2003). The control of glacial cycles in a fluvial system occurs at two different levels: Upstream Controls associated with Continental Interiors and Downstream Control in continental margins. At the upstream level, the relationship between variation of climatic regime and fluvial morphodynamics is usually associated with hydrological changes, on a broader scale, as a regional manifestation of a global pattern of atmospheric circulation or internal dynamics, with feedback mechanisms, while in downstream systems, the base level variations (i.e. sea-level) are the underlying drivers of erosion and incision processes.

The variations of drainage patterns in upstream systems relate to climate changes, and their association with hydrologic processes has been the focus of investigations since the publication of the earliest works on fluvial geomorphology. Patterns of marginal terrace construction in glacial periods (in contrast to incision processes in interglacial periods driven by changes in hydrological and sedimentary processes) also can produce more dramatic landforms. The climatic conditions not only affect the hydrologic regime (i.e. the flow increase or decrease in stream discharge) but also act on factors such as the vegetation coverage, weathering processes (which determines the sediment supply to the system) and the role of the sediment transport agents.

The clearest examples of geographical changes in drainage landform patterns controlled by climatic agents are the drying and desertification processes, associated with an increasingly arid climate condition. Most of the world's deserts present clear signals of paleo drainage systems related to humid periods in the past. The Sahara paleo drainage, as an example, has been investigated by McCauley *et al.*, 1986; Pachur and Kroepelin, 1987, Burke *et al.* 1989; Robinson *et al.* 2006; Ghoneim and El-Baz, 2007 among others. Beneath the mobile Eolian sand sheet, sand covers several paleo basins and channels varying from a few to hundreds of meters in width. Most of these systems have been established during the Messinian salinity crisis period (late Miocene, 5–6Ma - previously discussed) when the incision activity reached its maximum, followed by recurring activity during wet periods in the Quaternary. Paiollou *et al.* 2009, using radar images and digital terrain modeling, mapped the Kufrah and Sirt basins (Southern Libya) and concluded that the ancient Kufrah/Wadi Sahabi Rivers would have been active during wet periods (until the Holocene) showing diverse depth levels of incised channels, moreover, during the Eemian interglacial maximum, 125 k.y. ago. The Kufrah River system would have been more than 1200 km-long, comparable in size to the Egyptian Nile.

Studies showing direct links between climatic data and geomorphological analysis of paleodrainage systems in semi arid as well as desert environments are relatively rare (Maizels, 1990; Langping *et al.* 2009) since the cause-effect relationships between long term rainfall and morphodynamic of paleodrainages are not always direct. In fact, recent studies have mapped near-surface paleochannels in Namibia (Lancaster *et al,* 2000; Lancaster, 2002) indicating that during the late Quaternary, there was no significant changes in precipitation in the desert itself, but there were considerable variations with more frequent and/or longer duration discharges in ephemeral rivers, suggesting an increase of precipitation only in the upper part of the basin.

It is not only deep variations in climatologic regimes, leading to desertification and drastic opposite states of wet and dry periods, that can be responsible for considerable changes within a drainage system. In some cases, smaller scale/higher frequency variations in factors such as sediment supply, vegetation coverage or catastrophic floods can provoke considerable displacements and shifts in the fluvial systems, with or without the presence of tectonic amplification. The concepts of "River Metamorphosis" (Schumm, 1967 and 1985), "Avulsion" and "Cutoff" (Allen, 1965) and "stream piracy" (Pederson, 2001) have been proposed to describe such processes of channel migration into the watershed and have been widely studied as they are often associated with changes in climatic conditions (e.g. Hocke 1977; Werritty and Ferguson , 1980; Ashworth and Ferguson, 1986; Bridge, 2003).

Knox, (2000) in an extensive study on the Mississippi and Colorado River Valley and tributaries, described how magnitude and recurrence of paleofloods are linked to major climatic events during the Holocene and suggests that, even relatively modest, changes in climate can cause important flood episodes, and these changes in morphodynamic processes have often occurred abruptly at various time scales from decadal to millennial and even longer. Similar results were described by Benito *et al.* (2003) for the Iberian Peninsula, St.Laurent *et al.* (2001) Canada, Sih et al. (1985) and China among others.

Despite that fluvial system adjustments to new climatic and hydrological regimes are able to produce major responses at upstream levels, the effects of base level control in downstream levels (in response to Glacio-eustatic fluctuations) effectively can produce clear paleodrainage records in continental margins and coastal plains, and the significance of these characteristics can be very elucidating for understanding geomorphic processes associated with sea level changes. The paleodrainage characteristics is one of the main geoindicators of continental margin evolution over the Glacial periods when most of the world´s shelves were exposed to sub aerial incision during relative sea level low-stands, and the cross-shelf valleys were directly connected to the continental slope and deep marine environments. From a sequence stratigraphy point of view (see Van Wagoner *et al.* 1990), an understanding of how, when, and where incised valleys are excavated in response to sea level variations remains a topic of considerable interest since the basal surface of paleovalley fills is commonly taken to define the sequence boundary, the most commonly used bounding surface for the identification of allostratigraphic units (Tronqvist *et al.* 2005).

Models developed by Koss *et al.* (1994), Ashley and Sheridan (1994), Meijer (2001); Gutierrez *et al.* (2003), Weber *et al.* (2004) and Hori *et al.* (2002) Strong and Paola (2006) suggest that, despite continental shelves, paleovalleys can be correlated with the paleo-hydrological regime in which the channel was formed; they also can be associated with submerging processes in the incised valleys during marine transgressions, reflecting the interaction between the autogenic and allogenic forcings. Still, the relative and absolute effect of these mechanisms remains unclear (Wellner and Bartek, 2003).

A quite thorough review on the fluvial systems response to sea level variations can be found in Bloom and Tronquvist, 2000: According to the authors, fluvial responses to climate and sea-level change can be disaggregated into stratigraphic, morphological and sedimentological components: The stratigraphic responses are produced by aggradation due to increasing accumulation space, degradation as accumulation space decreases and lateral migration when there is no change in accumulation space and the fluvial systems in fluvial systems in subsiding basins. These are characterized by long-term net aggradation, punctuated by relatively short periods (103 - 104 years) of incision and/or lateral migration (in opposition to typical upstream valleys). Morphological responses reflect changes in channel system geometry and sedimentological responses are associated with changes along with the spatial distribution of depositional environments and proprieties of lithofacies.

3. ANALYTICAL TECHNIQUE - TRACING OLD WATERCOURSES

Paleodrainage systems are a valuable tool for modeling past environmental conditions and comprehending the geological and climatic changes that have occurred over various time scales, ranging from short-term seasons to long-term millions of years. However, identifying and characterizing these systems is typically an indirect process that often necessitates a thorough analysis and integration of data from multiple sources. Andersen (1962) offers the most extensive analysis of fundamental concepts about this subject. He has already identified the primary geological factors involved in creating maps of paleodrainage distribution. The author states that paleodrainage maps can be generated using many methods: 1) Create a map that shows the shape of the erosion surface at the bottom of the valley fill. 2) Create several cross-sections that intersect the trends of the valley. The valley axes are approximated by connecting consecutive low points of the valley that have been plotted on a map. 3) Create a map that shows the geological history of the surface at the bottom of the valley fill. The valley axes are drawn in the middle of areas where the oldest rocks are found. 4) Measure the thickness of the interval between the valley floor and a reference point. The valley axes are determined based on whether the valley floor is farthest from or closest to the reference point. 5) Create a map that shows the thickness of the valley fill. The valley axes are determined based on where the fill is the thickest. Understanding the paleodrainage patterns in a specific region enables geologists to determine the direction of ancient slopes, the local topography of the past, the general direction of the source area relative to the study area, the nature of the surface on which the valleys were formed, and whether there was any geological activity in the study area when the valleys were being carved.

The development of new methods in geomorphology to record and quantify past and recent geomorphic processes allied with the implementation of techniques of geoprocessing and Geographic Information Systems have permitted increased accuracy in reconstructions of past environments. This new "geomorphological" approach has also provided valuable insights into the formation and evolution of fluvial systems. Mapping topographical attributes, for example, can still be the simplest and most intuitive method to recognize some aspects of the ancient fluvial geomorphology and usually can reveal several characteristics of the fluvial system. In practice, however, the original local landform is rarely preserved after morphodynamic processes that caused the changes in the fluvial patterns, so the use of elevation data (obtained from different sources) should be accompanied by further geophysical investigations, in order to track aspects of the original landform configuration.

Examples of using of integrated methods based on multi parameter investigation of paleodrainage can be related to drying and desertification processes in which the original landform still lies only a few meters above the sand sheet, still remaining recognizable. Ghoneim & Elbaz (2007), based on SRTM elevation data obtained on a near-global scale (3 arcsec resolution) in combination with Radarsat-1 Synthetic Aperture Radar Images and Landsat TM satellite images, made a complete paleoenvironmental analysis of the Tushka basin between Egypt and Sudan (figure 3). Similar methods were used by Sultan *et al* (2004), Marinangeli (2005) Gaber *et al.* (2008) Blumberg *et al.*, (2004); Griffin, (2006); Pachur and Altmann, (2006) demonstrating good potential for recognizing topographic patterns of paleodrainage in semi arid or arid areas.

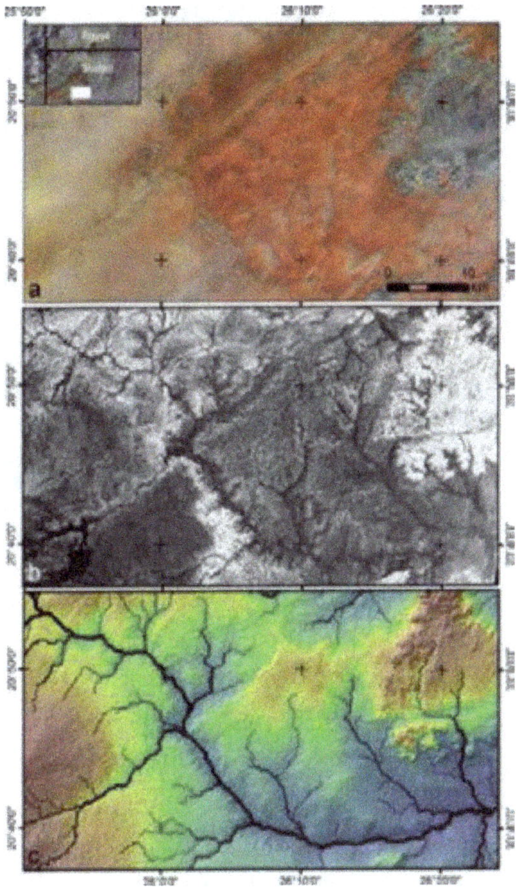

Fig. 3. (a) typical example of multi data INTEGRATED to characterize paleodrainage networks - Landsat ETM+ color composite (bands 7, 4, 2) covering an area in NW Sudan - showing the dominance of surface sand deposits. (b) Radarsat-1 image showing a distinct subsurface drainage network trending eastward; note the absence of the surface sand deposits. (c) The SRTM DEM (90 m) with the delineated drainage network overlain; note the good correspondence between the SRTM-generated channels and the Radarsat-1 subsurface channels (after Ghoneim & Elbaz, 2007).

In addition, topographical analysis combined with spectral information has been used to generate fascinating research in the recognition of paleodrainage in tropical/equatorial woods. The researchers Haywakawa et al. (2010) conducted an analysis of the ancient drainage complex of the Madeira River, a significant tributary of the Amazon. They successfully identified a complex network of abandoned channels resulting from avulsion processes that occurred during the Late Pleistocene-Holocene period. This discovery was made even though the area was densely covered by forests. (See figure 4 for visual representation.) Rossetti (2010) has achieved good results in comprehending the drainage rearrangement in the Amazonian lowlands by studying topographic and spectral data.

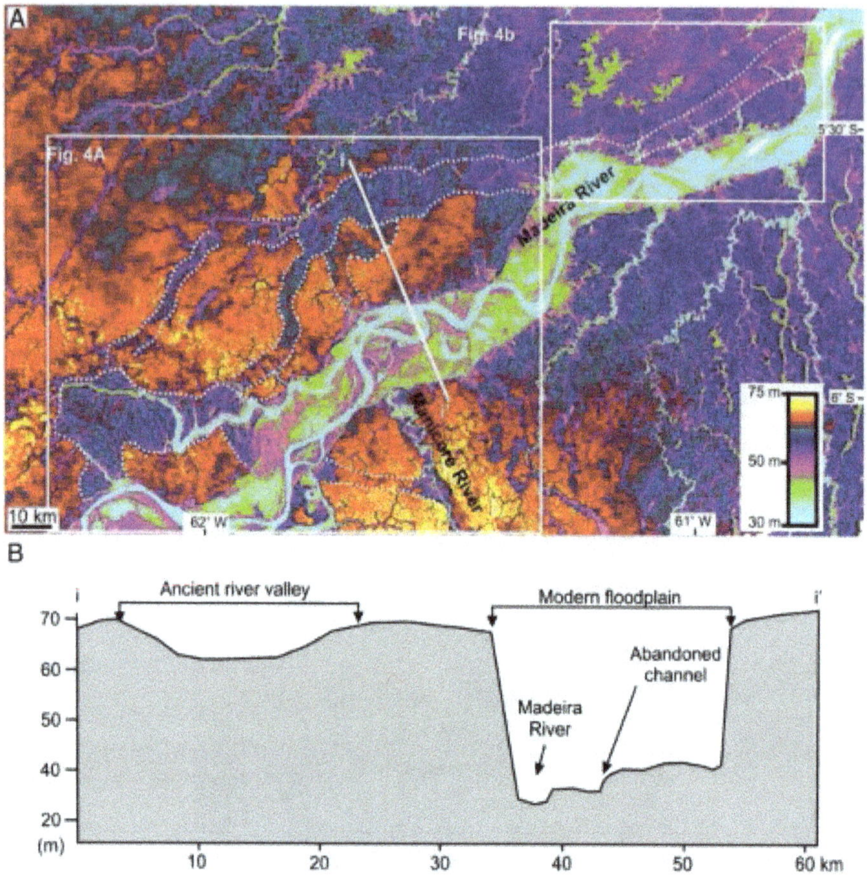

Fig. 4. DEM-SRTM with the regional view of the study area and the location of the paleodrainage network (dashed lines) detected in the Madeira-Manicoré Rivers. B) Topograhic profile (i–i¢) derived from DEM-SRTM. Observe that paleodrainage occurs in an area topographically higher than the modern Madeira valley (after Haywakawa *et al.* 2010).

In downstream systems, the subsurface mapping for locating and analyzing cut and fill structures associated with buried incised channels across continental shelves is probably one of the most widespread applications involving paleodrainage studies. Several studies have considered these factors as indicators of the paleogeography of the subaerial surface as well as the fluvial and hydrological regime during regressive events (see Reynaud *et al.*, 1999; Lericolais *et al.*, 2003; Tesson *et al.*, 2005; Meijer, 2002).

Despite the focus on direct analysis of channel incision surfaces and their relation to quaternary stratigraphy based on high resolution seismic data (e.g. Thorne, 1994; Lericolais *et al.*, 2001; Hori *et al.*, 2002; Nordfjord *et al.* 2005, Ashley and Sheridan, 1994; Zaitlin *et al.*, 1994), along with the fact that they are covered by marine sediments deposited during and after rises in sea levels (Correa, 1996;

Harris and O'Brien, 1996; Collina-Girard, 1999; Abreu and Cagliari, 2005; Finkl *et al.*, 2005) and the small number of studies characterizing the topographic expression of the paleo-drainage registers (Strong and Paola, 2008;), it is still possible to identify and classify the main paleo-valley axis in the surface topography.

Conti and Furtado 2009, having studied the paleodrainage registers of the southeastern Brazilian continental shelf, observed that the paleo-valley features present diverse topographical expressions and are related differently to the drainage system that generated them. Some of those identified by the Digital Terrain Model (DEM) were quite conspicuous and easily detected, while others could only be distinguished by using enhanced filters on the topography DEM; however, there is no direct relation between the size and dimension of the paleo-valleys and the shape, form, outflow or basin characteristics of the modern rivers.

A particularly interesting example of how the use of multi data in a Geographic Information System framework proved to be valuable in understanding the paleo drainage records is the case of the downstream of Ribeira de Iguape River – RIR (southeastern São Paulo state – Brazil). The analysis of remote sensing images (Landsat TM), in particular near infrared bands of the coastal plain of the river, reveal areas with high wetness index values indicating an abandoned channel feature with a divergent orientation in relation to the modern RIR position (figure 4a). A bathymetry digital terrain model (DTM) of the adjacent Continental Shelf mapped two channel-like structures in submarine topography; one clearly related to the present estuary of the RIR and another, more than 15 km southward, apparently connected to the paleochannel have been identified on satellite images (figure 4b). High-resolution seismic profiling data on the supposed point of convergence of these two features (satellite images and DTM) revealed the presence of a prominent cut and fill structure indicative of channel incision (figure 4c). The body of evidence points to two distinct positions of the RIR during two low-stand periods: one associated with the actual position of the river at the north and another one, probably older, diverging more than 30 km from the present estuary, headed seaward in a mostly direct path, forming the paleovalley features in the coastal plain and on the continental shelf.

Incised valleys are also often associated with large-scale sediment bypass zones that feed downdip into large sandy deltas, so the stratigraphic characterizing of these features can provide reliable information on the general characteristics of a paleofluvial regime (Dalrymple *et al.*, 1994). Different sedimentation models based on the fluvial and marine deposition during relative rises and falls of sea level can be constructed based on direct (biostratigraphic, isotopic, and geochemistry data) and indirect (seismic data) analysis of such submarine fans.

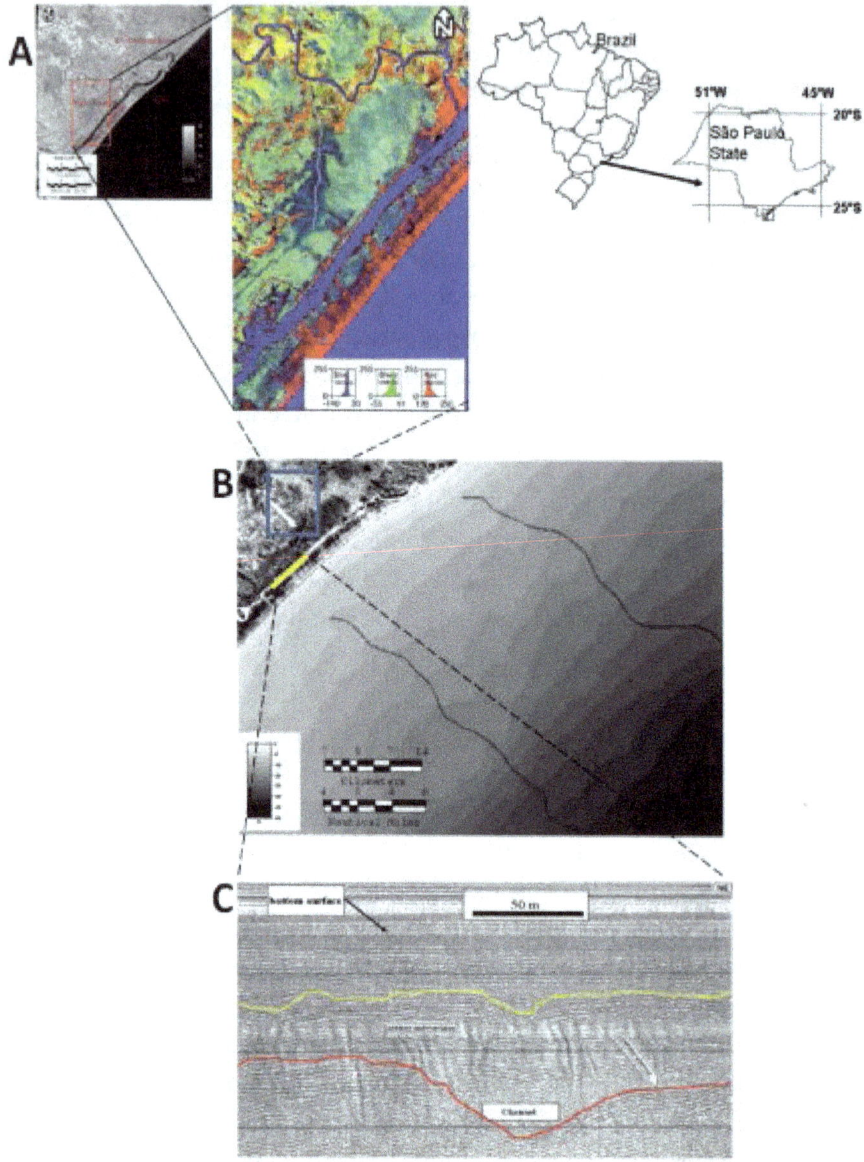

Fig. 5. Indicators of the paleo Ribeira de Iguape River – Southeastern Brazil. (a) Position of a wet lowlands indicated by vegetation indices from Landsat Images. (b) Digital Terrain Model from the continental shelf bathymetry indicating the presence of two paleovalleys (c) cut and fill structure in a high resolution seismic profile.

Examples in the Amazon (Mikkelsen *et al.* 1997 and Figueredo *et al.* 2009), Indus (Qayyum *et al.* 1997), Danube (Panin *et al.* 1998), and Pearl River (Pang *et al.* 2006) among others, have proven that the sediment load variability respond not only to climate changes that affect the river discharges but also to processes within the basin such as tectonic and autogenic processes.

Advancements in genetic analysis, namely in the field of biogeography and paleoenvironmental reconstructions, have significantly contributed to our understanding of the historical development of ancient drainage networks. Freshwater fish, specifically, have shown significant correlations between past river connections and phylogeographic patterns. This is because alterations in river patterns could have caused isolation and divergence among aquatic species, creating obstacles to the exchange of genetic material. In this context, a large number of studies have been suggesting even the influence of the paleodrainage on species evolution and vicariance processes as well as the other way around, resorting to biodiversity in order to reconstruct fluvial paleoenvironments (see Waters, 2001, Cook *et al.*, 2006; Ribeiro, 2006; Burridge *et al.* 2007; Flinstone *et al.* 2007).

4. FINAL REMARKS

There are numerous unresolved inquiries regarding the development of drainage systems in response to prolonged environmental changes. Some of these questions have been debated since the mid 20th century after the publication of major quantitative studies on dynamic fluvial geomorphology (e.g. Horton, 1945; Strahler, 1952; Shreve, 1969; Schumm, 1972; 1977 and 1987). Perhaps one of the most intriguing questions related to the evolution of fluvial systems is the relationship with base level low-stands. The models for incised valleys assume that relative sea-level fall produces an upstream propagating wave of rejuvenation, but how far reaching and how impacting are these processes, is still a matter for discussion. Other key aspects of paleohydrological and paleoenvironmental sciences remains unclear and uncertain, for example, the role of feedback effects on the climatic forcing (such as peak precipitation) to indirect (such as permafrost melting) and partial forcing (such as vegetation suppression) on the evolution of fluvial systems (Vandenberghe, 2003) and the temporal scales, time lapses and coupling of different stages of the geomorphic processes (Harvey, A.M., 2002). In fact, most of the models for reconstructing the morphology and hydrology dynamics in response to climate change (either in continental upstream or continental shelves downstreams) are quite incomplete in delivering quantitative descriptions and predictions, therefore, the development of integrated frameworks is fundamental to understanding such relationships.

In light of the ongoing progression of global warming, rapid urbanization, deforestation, and the increasing demand for hydroelectric energy, it is imperative that we prioritize the development of comprehensive and interdisciplinary research. This research should specifically focus on understanding the geomorphological and geological processes that shape landscapes, particularly fluvial systems. It is evident that the knowledge gained from studying natural drainage systems can be effectively applied to artificial drainage systems, regardless of their size or location. Currently, there are approximately one billion individuals residing in close proximity to rivers and channels across the globe (Palmer et al. 2008). To enhance the precision and dependability of comprehensive models and theories, it is imperative to establish a strong scientific foundation through the combination of empirical evidence of ancient river systems on both local and global scales, and theoretical models grounded in rigorous physical assumptions. This integration will enable the development of more accurate and reliable models and theories, thereby enhancing their applicability in forecasting, simulation, and planning endeavors.

REFERENCES

Al-Sulaimi, J., F.J. Khalaf, & A. Mukhopadhyay, (1997). Geomorphological Analysis of Paleo Drainage Systems and Their Environmental Implications in the Desert of Kuwait. Environ. Geol.Berlin 29(1/2):94-111.

Andres, W., Reis, J. & Seegar, M. (2002): Pre- Holocene sediments in the Barranco de las Lenas, Central Ebro Basin, Spain, as indicators for climate-induced fluvial activities. Quater ary International 93–94, 65–72.\

Ashley, G. M. & Sheridan, R. E. (1994) Depositional model for valley fills on a passive continental margin. In: Dalrymple, R. W.; Boyd, R.; Zaitlin, B. A. (Ed.). Incised-valley systems: origin and sedimentary sequences. Spec. Publs Soc. econ. Paleont. Miner.,Tulsa, v. 51, p 285-302,.

Ashworth, P.J. & Ferguson,R.I. (1986) Interrelationships of channel processes, changes and sediments in a proglacial braided river. Geografiska Annaler, 68A (1986), pp. 361– 371

Blum, M. D., & Tornqvist, T. E. (2000). Fluvial responses to climate and sea-level change: a review and look forward. Sedimentology, 47(s1), 2-48.

Branca, S., & Ferrara, V. (2001). An example of river pattern evolution produced during the lateral growth of a central polygenic volcano: the case of the Alcantara river system, Mt Etna ž Italy /. Catena, 45/2. 85-102

Bridge, J.S., (2003). Rivers and Floodplains: Forms, Processes, and Sedimentary Record, Blackwell, Oxford, 491 pp.

Brookfield, M. E. (1998). The evolution of the great river systems of southern Asia during the Cenozoic India-Asia collision: rivers draining southwards. Geomorphology, 22, 285-312.

Burbank, D. W., & R. S. Anderson (2001), Tectonic Geomorphology, 1st ed. Blackwell Sci., Malden, Mass

Burke, Kevin, & Welles, G.L., (1989), Trans-African drainage system of the Sahara: Was it the Nile?: Geology, v. 17. 743-747

Clark, M. K. (2004). Surface uplift, tectonics, and erosion of eastern Tibet from large-scale drainage patterns. Tectonics, 23(1), 1-21.

Clift, P. D., & Blusztajn, J. (2005). Reorganization of the western Himalayan river system after five million years ago. Nature, 438(7070), 1001-3

Collina-Girard, J. (1999) Observation en plongee de replats d'erosion eustatique a l'ile d'Elbe (Italie) et a Marie-Galante (Antilles): une sequence bathymetrique mondiale. C.r. Acad. Sci, . Paris, v. 328. 823–829,

Cook B.D., Baker A.M. & Page T.J. (2006) Biogeographic history of an Australian freshwater shrimp, Paratya australiensis (Atyidae): the role of life history transition in phylogeographic diversification. Molecular Ecology, 15, 1083–1093.

Coutellier, V. & Stanley, D.J. (1987). Late Quaternary stratig-raphy and paleogeography of the eastern Nile delta, Egypt. Marine Geology, 77, 257-275.

Dalrymple, R. W.; Boyd, R & Zaitlin, B. A. (1994) History of research, types and internal organization of incised-valley systems: introduction to the volume. In: DALRYMPLE, R. W.; BOYD, R.; ZAITLIN, B.A. (Ed.). Incised-valley systems: Origin and sedimentary sequences. Spec. Publs Soc. econ. Paleont. Miner.,Tulsa, n. 51, 3-10.

Davis, W. M. (1896) The Seine, the Meuse and the Moselle. Ann. de géographie, 25-49

Diaz de Gamero, M. L., (1996), The changing course of the Orinoco River during the Neogene: A review: Palaeogeography, Palaeoclimatology, Palaeoecology, v. 123. 385–402

Dollar E.S.J. (2004) Fluvial geomorphology. *Progress in Physical Geography*, 28, 405-450

Downing, K. F., & Lindsay, E. H. (2005). Relationship of Chitarwata Formation paleodrainage and Paleoenvironments to Himalayan Tectonics and Indus River Paleogeography. Palaeontologia Electronica, 8(1), 1-12.

Ely LL, Enzel Y, Baker VR & Cayan DR. (1993). A 5000-year record of extreme floods and climate change in the southwestern United States. Science 262: 410–412.

Figueiredo, J., Hoorn, C., Ven, P. & Soares, E. (2009). Late Miocene onset of the Amazon River and the Amazon deep-sea fan: Evidence from the Foz do Amazonas Basin. *Geology*, *37*(7), 619-622.

Finnegan, N. J., Hallet, B., Montgomery, D. R., Zeitler, P. K., Stone, J. O. & Anders, A M.(2008). Coupling of rock uplift and river incision in the Namche Barwa-Gyala Peri massif, Tibet. *Geological Society of America Bulletin*, *120* (1-2), 142-155.

Finston, T. L., Johnson, M. S., Humphreys, W. F., Eberhard, S. M., & Halse, S. A. (2007). Cryptic speciation in two widespread subterranean amphipod genera reflects historical drainage patterns in an ancient landscape. *Molecular ecology*, *16*(2), 355-65.

Foucault & D.J Stanley (1989), Late Quaternary palaeoclimatic oscillation in East Africa recorded by heavy minerals in the Nile delta. *Nature*, 339 . 44–4

Frumkin, A., Greenbaum, N. & Schick, A.P. (1998). Paleohydrology of the Northern Negev: comparative evaluationof two catchments. In: Issar, A.S., Brown, N., eds. Water, environment and society in times of climatic change. Kluwer, the Netherlands, 97– 111.

Gautier, F., Clauzon, G., Suc, J.P., Cravatte, J. & Violanti, D., (1994). Age and duration of the Messinian salinity crisis. C.R. Acad. Sci., Paris (IIA) 318, 1103–1109

Ghoneim, E. & Elbaz, F. (2007). The application of radar topographic data to mapping of a mega-paleodrainage in the Eastern Sahara. *Journal of Arid Environments*, *69*(4), 658- 675. doi: 10.1016/j.jaridenv.2006.11.018.

Ghoneim, E. & Elbaz, F. (2007). The application of radar topographic data to mapping of a mega-paleodrainage in the Eastern Sahara. *Journal of Arid Environments*, *69*(4), 658- 675. .

Gibbard, P.L. & Lewin, J. (2002) Climate and related controls on interglacial fluvial sedimentation in lowland Britain. *Sedimentary Geology*, 151 . 187–210

Goodbred, J.S.L. (2003) Response of the Ganges dispersal system to climate change: a source-to-sink view since the last interstade. *Sedimentary Geology*, 162 . 83–104.

Goudie, A. (2005). The drainage of Africa since the Cretaceous. Geomorphology, 67(3-4), 437-456.

Gregory-Wodzicky K. M.(2000) Uplift history of the Central and Northern Andes: A review. GSA Bull.;112:1091-1105

Hayakawa, E. H., Rossetti, D. F., & Valeriano, M. M. (2010). Applying DEM-SRTM for reconstructing a late Quaternary paleodrainage in Amazonia. *Earth and Planetary Science Letters*, *297*(1-2), 262-270.

Hodges, K.V. (2000), Tectonics of the Himalaya and southern Tibet from two perspectives: Geological Society of America Bulletin, v. 112. 324–350

Hoorn, C., Guerrero, J., Sarmiento, G. A., Lorente, M. A., Vries-laboratorium, C. H. H. D., & Amsterdam, S. M., (1995). Andean tectonics as a cause for changing drainage patterns in Miocene northern South America. Geology. v. 23. 237-240

Hoorn, C., Wesselingh, F. P., Steege, H., Bermudez, M. A, Mora, A, Sevink, J. (2010). Amazonia through time: Andean uplift, climate change, landscape evolution, and biodiversity. *Science (New York, N.Y.)*, *330*(6006), 927-31.

Horton, R. E. (1932). Drainage basin characteristics. Transactions of the American Geophysical Union 13:350-361.

Horton, R. E. (1945). Erosional development of streams and their drainage basins: hydrophysical approach to quantitative morphology. Geological Society of America Bulletin 56:275-370.

Issawi B. & Mccauley J.F. (1992) - The Cenozoic rivers of Egypt: the Nile problem. In Adams B. & Friedman R., Editors, The Followers of Horus: studies dedicated to Michael Allen Hoffman 1944-1990. Egyptian Studies Association Pubblication, 2, Oxbow Press, Oxford, 121-138.

Knox, J. C. (2000). Sensitivity of modern and Holocene floods to climate change. *Quaternary Science Reviews*, *19*, 439-457.

Koons, P.O., Zeitler, P.K., Chamberlain, C.P., Craw, D., & Meltzer, A.S., (2002), Mechanical links between erosion and metamorphism in Nanga Parbat Pakistan Himalaya: American Journal of Science, v. 302. 749–773.

Lancaster, N., Schaber, G. G., & Teller, J. T. (2000). Orbital Radar Studies of Paleodrainages in the Central Namib Desert. *Area*, *4257*(99). 216-225

Langping, Y.D., Huang D., Zhaoshuai, G. Xu Q & Wang Xi. (2009) Drainage Evolution of Qiaojia-Xinshizhen Section of Jinsha River.纪研究,,29(2): 327-333

Leopold, L. B., & T. Maddock, Jr. (1953). The hydraulic geometry of stream channels and some physiographic implications. U.S. Geological Survey Professional Paper 252. 57.

Maizels, J. (2009) Raised channel systems as indicators of palaeohydrologic change: a case study from Oman. Palaeogeogr. Palaeoclimatol.Palaeoecol., 76, 241–277.

Marston, R. A., Girel, J., Pautou, G., Piegay, H., & Bravard, P. (1995). Channel metamorphosis , floodplain disturbance , and vegetation development : Ain River , France. *Water Resources*, *13*, 121-131.

McCauley,F.. Breed C.S & Schaber, G.G.*et al.* *(1986)* Paleodrainages of Eastern Saharathe radar rivers revisited (SIR-A/B implications for a mid-tertiary trans-African drainage system). *IEEE Trans. Geosci. Remote Sens.*, GE-24 4. 624–648.

Mikkelsen, N., Maslin, M., Giraudeau, J., & Showers, W., (1997), Biostratigraphy and sedimenta- tion rates of the Amazon Fan, in Flood, R.D., et al., Proceedings of the Ocean Drilling Pro- gram, Scientifi c results, Volume 155: College Station, Texas, Ocean Drilling Program. 577–594

Najman, Y., Garzanti, E., Scienza, P., Bickle, M., & Stix, J. (2003). Early-Middle Miocene paleodrainage and tectonics in the Pakistan Himalaya. October, (10), 1265-1277.

Nugent, C. (1990). The Zambezi River: tectonism, climatic change and drainage evolution. Palaeogeography Palaeoclimatology Palaeoecology, 78, 55-69.

Oilier, C. D. (1995). Tectonics and landscape evolution in southeast Australia. *Main*, *12*, 37- 44.

Pachur, H.-J., & Kroepelin, Stefan, (1987), "Wadi Howar: Paleoclimatic evidence for an extinct river system in the southeastern sahara," *Science* v. 237. 298-300.

Paillou, P., Schuster, M., Tooth, S., Farr, T., Rosenqvist, A., Lopez, S., et al. (2009). Mapping of a major paleodrainage system in eastern Libya using orbital imaging radar: The Kufrah River. *Earth and Planetary Science Letters*, *277*(3-4), 327-333.

Pang X, Chen C M, Wu M S, et al. (2006) The Pearl River deep-water fan systems and significant geological events. Adv Earth Sci, 2006, 21(8): 7—14

Panin N. (1996). Impact of global changes on geo-environmental and coastal zone state of the Black Sea. Danube Delta-Black Sea System under Global Change Impact. In: Malit A., Gomoiu M.-T and Panin N. (eds.), Geo-Eco-Marina 1. 1–6.

Pederson, D.T., (2004). Stream piracy revisited: a groundwater sap- ping solution. GSA Today 11 (9), 4–10 Proceedings of the 2004 Envisat & ERS Symposium (ESA SP- 572). 6-10 September 2004, Salzburg, Austria. Edited by H. Lacoste and L. Ouwehand. Published on CD-Rom., #173.1

Penck, A., (1910). Versuch einer Klimaklassifikationauf physiographische Grundlage, Preussische Akademie der Wissenschaften. Sitz der Phys.-Math., Kl. 12, 236–246.

Penck, A. (1927) "Geography Among the Earth Sciences" in *Proceesings of the American Philosophical Society* 66. 621–644

Potter, P. (1997). The Mesozoic and Cenozoic paleodrainage of South America: a natural history. Journal of South American Earth Sciences, 10(5-6), 331-344.

Qayyum, R.D. Lawrence & A.R. Niem (1997) Discovery of the Palaeo-Indus delta–fan complex. *Journal of the Geological Society (London)*, 154. 753–75

Ribeiro, A. C. (2006). Tectonic history and the biogeography of the freshwater fishes from the coastal drainages of eastern Brazil : an example of faunal evolution associated with a divergent continental margin. *America*, *4*(2), 225-246.

Robinson, C., El-Baz, F., Al-Saud, T.& Jeon, S., (2006). Use of radar data to delineate palaeodrainage leading to the Kufra oasis in the Eastern Sahara. Journal of African Earth Sciences 44, 229–240

Rogers, K. V. (2000). Tectonics of the Himalaya and southern Tibet from two perspectives. GSA Bulletin, 112(3), 324-350.

Rossetti, D. D. F. (2010). Multiple remote sensing techniques as a tool for reconstructing late Quaternary drainage in the Amazon lowland. *Earth Surface Processes and Landforms*, *35*(10), 1234-1239.

Saint-Laurent, C. Couture & E. McNeil (2001), Spatio-temporal analysis of floods of the Saint-Francois drainage basin, Quebec, Canada. *Environments*, 29 .74–90

Sanderson, D.C.W., Bishop, Stark M.T. & Spencer, J.Q. (2003) Luminescence dating of anthropogenically reset canal sediments from Angkor Borei, Mekong delta, Cambodia. *Quaternary Science Reviews*, 22. 1111–1121

Schumm S.A. (1977) *The Fluvial System*. Wiley, New York

Schumm, S.A. Mosley M.P. and Weaver W.E. (1987) Experimental Fluvial Geomorphology, Wiley, New York .

Schumm, S.A., (1967). Meander wavelength of alluvial rivers. Science, 157(3796): 1549-1550. Shahin, M., (1985): *Hydrology of the Nile Basin.* Elsevier, 575 pp.

Shih F. Shih, Y. Yi & Han, M. (1985) Investigation and verification of extraordinary large floods of the Yellow River in China Proceedings of the U.S. China Bilateral Symposium on the Analysis of Extraordinary Flood Events. Nanjing, China

Shreve. R. L. (1969) Stream Lengths and Basin Areas in Topologically Random Stream Networks. Jour. Geol., 77(4):397–414,.

Simpson, G. (2004). Role of river incision in enhancing deformation. *Society*, (4), 341-344.

Smith, R.D. & Swanson F. J, (1987) Sediment routing in a small drainage basin in the blast zone at Mount St. Helens, Washington, U.S.A., Geomorphology, Volume 1, Issue 1, 1-13,

Stankiewicz, J., & Dewit, M. (2006). A proposed drainage evolution model for Central Africa—Did the Congo flow east? *Journal of African Earth Sciences*, *44*(1), 75-84.

Strong , N., & Paola, C. (2008). Valleys That Never Were: Time Surfaces Versus Stratigraphic Surfaces. *Journal of Sedimentary Research*, *78*(8), 579-593.

Strong, N., & Paola, C., (2006), Fluvial landscapes and stratigraphy in a flume: The Sedimentary Record, v. 4, (2). 4–7.

Sultan, M., Manocha, N., Becker, R. & Sturchio, N., (2004), Paleodrainage networks recharging the Nubian Aquifer Dakhla and ufra sub-basins revealed from SIR-C and SRTM Data. EOS Transactions, 85(17), Jt. Assem. Supplement, H31C-03.

Thomas, D.S.G. & Shaw, P.A. (1991). The Kalahari environment. Cambridge University Press, U.K., 284 pp.

Twidale, C. (2004). River patterns and their meaning. Earth-Science Reviews, 67(3-4), 159- 218.

Van Wagoner, J. C., R. M. Mitchum, K. M. Campion, & V. D. Rahmanian, (1990), Siliciclastic sequence stratigraphy in well logs, cores, and outcrops: Concepts for high- resolution correlation of time and facies: AAPG Methods in Exploration 7, 55 p.

Vandenberghe, J. (2003). Climate forcing of fluvial system development: an evolution of ideas. *Quaternary Science Reviews*, *22*(20), 2053-2060.

Vandenberghe, J. (2003). Climate forcing of fluvial system development: an evolution of ideas. *Quaternary Science Reviews*, *22*(20), 2053-2060.

Wallbrink, D.E. Walling &. He, Q. (2002) Radionuclide measurements using HpGe gamma spectrometry, F. Zapata, Editor, *Handbook for the Assessment of Soil Erosion and Sedimentation Using Environmental Radionuclides*, Kluwer, Dordrecht. 67–96.

Waters (2001) Genes meet Geology: Fish Phylogeographic Pattern Reflects Ancient, Rather than modern, Drainage Connections. palaeogeography, palaeoclimatology, palaeoecology 76, (3-4), 241-277.

Waters JM, Craw D, Youngson JH & Wallis GP (2001) Genes meet geology: fish phylogeographic pattern reflects ancient, rather than modern, drainage connections. Evolution, 55, 1844–1851 Wellner, R. W., & Bartek, L. R. (2003). The Effect of Sea Level, Climate, and Shelf Physiography on the Development of Incised-Valley Complexes: A Modern Example From the East China Sea. Journal of Sedimentary Research, 73(6), 926-940.

Werritty, A. & R.I. Ferguson, R.I (1980) Pattern changes in a Scottish braided river over 1, 30 and 200 years,In. R.A. Cullingford, D.A. Davidson, J. Lewin, Editors , *Timescales in Geomorphology*, Wiley, Chichester. 53–68.

Wescott, W.A. 1993 Geomorphic thresholds and complex response of fluvial systems – some implications for sequence stratigraphy. Am. Assoc. Petrol. Geol. Bull., 77, 1208- 1218.

Wheeler, H.E. (1958) Time-stratigraphy. *Am. Assoc. Petrol. Geol. Bull.*, 42 . 1047–1063

Zaitlin, B. A.; Dalrymple, R. W. & Boyd, R. (1994) The stratigraphic organization of incised- VALLEY systems associated with relative sea-level change. In: Dalrymple, R.W.; Boyd, R. B.A. ZAITLIN, B. A. (Ed.). Incised-valley systems: Origin and sedimentary sequences Spec. Publs Soc. econ. Paleont. Miner., Tulsa,n. 51,. 45–62,.

Zeitler, P.K., Meltzer, A.S., Koons, P.O., Craw, D., Hallet, B., Chamberlain, C.P., Kidd, W.S.F., Park, S.K., Seeber, L., Bishop, M., & Shroder, J., (2001), Erosion, Himalayan dynamics and the geomorphology of metamorphism: GSA Today, v. 11,. 4–9.

Zhang, L.P. Zhou & S.Y. Yue, (2003) Dating fluvial sediments by optically stimulated luminescence: selection of equivalent doses for age calculation. *Quaternary Science Reviews*, 22 . 1123–1129

02 SUSTAINABLE DRAINAGE SYSTEMS: A COMPREHENSIVE APPROACH THAT COMBINES HYDRAULIC ENGINEERING DESIGN, URBAN LAND CONTROL, AND RIVER MANAGEMENT

1. INTRODUCTION: AN OVERVIEW

Floods are natural and cyclical occurrences that provide a significant ecological function. However, when they affect urban areas, they result in various types of losses. Urban expansion exacerbates urban floods. The alteration of land use through deforestation and the expansion of impermeable surfaces results in an increase in both the quantity and speed of runoff. Urbanization has become increasingly intense in recent years, although the issues of floods and drainage have been associated with city development since ancient times. Drainage systems are integral components of urban infrastructure and play a crucial role in urban living. In the event of a drainage system failure, cities are at risk of experiencing floods, potential environmental deterioration, sanitation and health issues, and disruption of city services. metropolitan rivers have historically served several purposes, including providing water supply, defending metropolitan districts, delivering goods, and conveying garbage.

Therefore, there exists a paradoxical relationship between water and cities: water is an essential component for urban life, yet urbanization often lacks proper planning and necessary infrastructure, resulting in the deterioration of both urban spaces and water resources. A fascinating historical record exemplifies the issue of urban land encroachment. Giovani Fontana, an architect of the 16th Century, conducted a study on the Tiber River flood that occurred during

Christmas of 1598 in Rome (Biswas, 1970). Fontana's findings indicated that the significant repercussions of the flood were linked to the settlement of riparian regions around the confluences of the Tiber River with various tributaries and channels. Additionally, the lack of awareness among the residents who built their homes in such locations contributed to the severity of the consequences. This situation bears a striking resemblance to contemporary issues: inadequate urban planning and regulation, insufficient environmental education, and the absence of a comprehensive framework to integrate technological and socio-economic factors. Fontana's primary proposal for flood control in Rome involved expanding the Tiber River to enhance overall flow conditions. This approach aimed to quickly direct floodwaters downstream to a secure release point, following a traditional perspective.

With the expansion of cities, particularly throughout the Industrial Era, the challenges of urbanization intensified, leading to a rise in both the scale and frequency of urban floods. The conventional method used for drainage systems, which were crucial for maintaining sanitation during the early stages of urban expansion, by transporting stormwater and wastewater, became unsustainable. There was an increase in the production of flow, and solutions that just addressed the end of the process tended to simply shift problems to the next stage. Over the past few decades, various methods have been developed to more accurately match flow patterns in both space and time. However, it is not solely the hydraulic characteristics that are crucial. Technical measures are not independent. It is important to take a comprehensive approach to the city's water management, and sustainable drainage systems must take into account urban revitalization and river rehabilitation. These systems should aim to improve the quality of life in communities, involve the participation of residents, and establish institutional arrangements that ensure the acceptance, support, and long-term implementation of these solutions.

2. HISTORICAL CONTEXT AND BACKGROUND OF URBAN FLOODS AND DRAINAGE SOLUTIONS

Various ancient civilizations demonstrated meticulousness in the construction of urban drainage systems, effectively achieving the goals of rainwater collection, flood prevention, and garbage conveyance. Urban drainage systems experienced tremendous advancements throughout the Roman Empire era. The city of Rome, situated in the Lazio region and built on the banks of the Tiber River, had significant concerns regarding the mitigation of urban flooding and the drainage of low-lying areas. In order to address the requirements of urban drainage, an intricate system of open canals and subterranean pipes was built. This method was also utilized for the transportation of individuals' refuse from their residential zones. (Burian and Edwards, 2002).

In the Middle Ages, urban centers experienced significant deterioration, leading people to prefer living in sparsely populated settlements located in rural areas near rivers. These towns had little regard for urban drainage. Sanitation standards declined following the fall of the Roman Empire, and wastewaters were disposed of and transported solely through surface drains and streets without discrimination (Chocat *et al.*, 2001). Later, when cities started to grow significantly again, in the Industrial Era, urban drainage found itself regretted to a second plane. The industrial city grew with very few guidelines. The Liberalism influenced urban growth and there was a certain lack of control on the public perspective for city development (Benevolo, 2001). Sanitation, then, became a great problem and inadequate waste disposal led to several sort of diseases and deterioration of public health. The role of urban drainage became very important in helping to solve this problem and, more than often, it was important to fast collect, conduct and dispose securely stormwater and wastewater. Focus was driven to improve conveyance and this was the main goal of urban drainage, until some decades ago. However, considering the fast urban growth of the last two centuries, and the fact that the world population profile is changing from rural to urban, it became hard to simply look at urban drainage and propose channel corrections, rectifications and other similar sort of interventions. Canalisation could not answer for all urban flood problems and, in fact, this isolated action, in a local approach, was responsible for transferring problems more than solving them. The increasing flood problems that the cities were forced to face showed the unsustainability of the traditional urban drainage conception and new solutions started to be researched.

A sustainable approach for drainage systems became an important challenge to be dealt with. Drainage engineers became aware that the existing infrastructure was overloaded.

Focus on the consequences of the urbanisation process, that is, the increase of flow generation, which concentrates on storm drains, should be changed. Source control, acting on the causes of flooding and focusing on storage and infiltration measures, emerged as a new option at the end of the 1970s (Andoh & Iwugo, 2002).

An integrated approach, considering the watershed as the planning unit, may be considered the initial basis for a sustainable system design. The design of an urban drainage system integrated with city development, aiming to reduce impacts on the hydrological cycle, acting on infiltration processes and allowing detention on artificial urban reservoirs, joining concerns, restrictions and synergies from Hydraulic Engineering and Urbanism appears as a fundamental option to treat urban floods. Besides the quantity aspects, the water quality became also a main issue and waste waters and solid waste disposal became matters to be treated together. The first flush and the washing of the catchment also introduced a new perception related to the diffuse pollution. At last, and in a complementarily way, rain water harvesting appears as an opportunity to increase water resources availability in urban environments. Several different conceptions have been proposed in the last decades, with some minor differences among them. All of them, however, tend to consider those questions in an integrated way, trying to rescue natural characteristics of the hydrological cycle, while adding value to the city itself.

Coffman et al. (1998) proposed a design concept of Low Impact Development (LID). LID design adopts a set of procedures that try to understand and reproduce hydrological behaviour prior to urbanisation. In this context, the use of functional landscapes appear as useful elements in the urban mesh, in order to allow the recovering of infiltration and detention characteristics of the natural watershed. It is a change in the traditional design concepts, moving towards a site design that mimics natural watershed hydrological functions, involving volume, discharge, recharge and frequency. The main principles of this approach may be briefly described by the following points:

- Minimise runoff, acting on impervious rates reduction and maintaining green areas;
- Preserve concentration times of pre-development, by increasing flow paths and surface roughness;
- Use of retention reservoir for peak discharge control and improve water quality;
- Use of additional detention reservoirs to prevent flooding, if necessary.

In a similar way, another early trend in the drainage system design evolution involved the use of stormwater Best Management Practices (BMP). The term Best Management Practices is frequently used in the USA and Canada and its origin is related with pollution control in the field of industrial wastewaters. Later, it was also referred as a possibility of nonpoint source pollution control and then associated with stormwater management. This way, stormwater BMPs are supposed to work in a distributed way over the watershed, integrating water quantity and water quality control aiming to mitigate effects generated by land use changes, with optimised costs. BMPs are designed to reduce stormwater volume, peak flows, and nonpoint source pollution through infiltration, filtration, biological or chemical processes, retention, and detention. They also may be classified into structural, when referring to installed devices and engineering solutions, or non-structural, when related to procedures changes, like limitations on landscaping practices (US EPA, 2004).

LID and BMP are very often used together and may complement each other. Batista *et al.* (2005), in Brazil, consolidated the concepts of Compensatory Techniques in urban drainage design, which meant the introduction of several different measures, focusing on infiltration and storage capacity, with the aim of compensating urban impacts on the hydrological cycle.

Another possibility of improving urban drainage solutions concerns the Sustainable Urban Drainage System (SUDS) concept. In this case, the ideals of sustainable development are included in the drainage system design process, that is, impacts on the watershed due to drainage solutions may not be transferred in space or time. Moreover, besides contributing to sustainable development, drainage systems can be developed to improve urban design, managing environmental risks and enhancing built environment. SUDS objectives account both for reducing quantity and quality problems and maximising amenities and biodiversity opportunities, which form the three way concept: quantity – quality – amenity & biodiversity. All of them have to be managed collectively and the desired solution appears in the interface of these three objectives (CIRIA, 2007). The philosophy of SUDS, similar to LID, is also to replicate, as well as possible, the natural conditions of pre-development site.

The key elements for a more sustainable drainage system consider to:

- Manage runoff volumes and rates, reducing the impact of urbanisation on flooding;
- Encourage natural groundwater recharge (where appropriate);
- Protect or enhance water quality;
- Enhance amenity and aesthetic value of developed areas;
- Provide a habitat for wildlife in urban areas, creating opportunities for biodiversity enhancement;
- Meet the environmental and the local community needs.

The continuous evolution of all these concepts and the seek for new urban drainage system solution led also to the Water Sensitive Urban Design (WSUD) concept, initially developed in Australia.

Wong (2006) states that the definition of WSUD appears to be confusing among practioners because of its wide range of applications. WSUD tries to integrate social and physical sciences in a holistic management proposition for urban waters.

Langenbach et al. (2008) define WSUD as the "interdisciplinary cooperation of water management, urban design and landscape architecture which considers all parts of the urban water cycle, combines water management function and urban design approaches and facilitates synergies for the ecological, economical, social and cultural sustainability."

According to Wong (2006): "WSUD brings 'sensitivity to water' into urban design. The words 'water sensitive' define a new paradigm in integrated urban water cycle management that combines the various disciplines of engineering and environmental sciences associated with the provision of water services, including the protection of aquatic environments in urban areas. Community values and aspirations of urban places necessarily govern urban design decisions and therefore water management practices".

WSUD is centred on integration at a number of levels (*ibid*):

- The integrated management of potable water, wastewater and stormwater;
- The integration of the urban water management from the individual allotment scale to the regional scale;
- The integration of sustainable urban water management with building architecture and landscaping;
- The integration of structural and non-structural sustainable urban water management initiatives.

The WSUD concept is characterized by the integration of urban water uses across various spatial scales, involving multiple knowledge areas such as hydraulic engineering, urbanism, architecture, social sciences, and economy. Its aim is to preserve the natural environment and enhance the built environment, within a participative framework that emphasizes the role of communities. The institutional arrangements play a crucial role in effectively managing this process.

Moreover, the prospect of undertaking initiatives to revitalize urban rivers, or more optimistically, to facilitate the restoration of urban rivers, is also emerging as a novel opportunity. River revitalisation typically involves strategies for the constructed surroundings, aiming to reintegrate it with the urban environment, without necessarily restoring natural patterns. The concept of river rehabilitation aims to incorporate the hydrology and morphology of the river, the hydraulic risks related to flood control, water quality, and the ecological condition of the river. Managing these duties in urban contexts is highly challenging due to numerous constraints that have accumulated over time. River canalization, the disconnection of flood plains, limited availability of open areas, the presence of combined sewers (or the frequent unregulated discharge of wastewater, particularly in developing countries), social constraints, and other challenges hinder the process of river restoration. One potential approach in urban development is to establish a clear vision for the river and strategize on how to seamlessly incorporate it into the surrounding infrastructure, striving to make progress in that direction.

Gusmaroli et al. (2011) propose the adoption of an ecosystem approach, in order to supplement or replace the concept of Waterfront Design. The primary objective of the Waterfront Design is to restore the connection between the river and the city at the point where they intersect. Advancing allows us to present the idea of rehabilitating the river as a means of enhancing the environment. We view the city as a dynamic organism that can adapt and change, even partially, to restore the natural characteristics of the waterways. It is a difficult task to discover methods to restore rivers to their natural state and reconsider urban development as a consequence.

3. INTERFACE BETWEEN RAIN WATERS AND THE CITY

Urbanization dramatically alters the equilibrium of the natural water balance. The removal of vegetation and its replacement with impermeable surfaces decreases the ability for water to infiltrate, resulting in an increase in the volume of surface runoff. In addition, the natural ability to retain water is decreased and the flow of water across controlled urban surfaces is accelerated. Generally, even when urbanization is carried out according to planning requirements, the surface volumes are larger. If urbanization fails to incorporate sustainable patterns, the peak flow exceeds the natural flow and the peak time occurs prematurely.

Unregulated urban expansion, particularly in emerging nations, where rapid industrialization has resulted in rapid city growth, often results in the encroachment of natural floodplains and even riverbanks, as depicted in figure 1. This exacerbates the issue of urban floods, as the areas required for flood overbank flows are now taken up by residential buildings, roads, and infrastructure. During such circumstances, floods have a tendency to expand over vast distances, seeking space, and impacting various aspects of urban life such as sanitation, health, traffic, housing, resulting in significant losses. As urbanization reduces the available flood space, the rushing waters seek alternative routes, causing formerly unaffected areas to become inundated.

Fig. 1. Riverbanks occupation in the metropolitan area of Rio de Janeiro City, Brazil (Photo Miguez, 2010).

Typically, following the initial effects of urban development, such as changes in land use and the occurrence of initial floods, the construction of drainage canals becomes a common consequence. This occurs at both the local level, with the installation of small-scale drainage systems, and at the regional level, with the implementation of large-scale drainage canal projects. Canalization operations are often associated with the construction of roadways and the establishment of regular city grids. Canalization, as previously said, has a limited perspective on the issue and only addresses flooding in specific areas.

Typically, the process of occupying a watershed begins at the downstream locations, specifically in the lower and level regions. The impermeability and urbanization of riverine areas necessitated the implementation of canalization as a means of flood control and urban design in these regions, exacerbating flooding at the basin outfall. As urbanization continues, the lands upstream are also being populated, following the same patterns as the ones downstream. Therefore, as these newly developed regions begin to encroach against the riverine areas and experience flooding, new channels are constructed, resulting in the older downstream districts, where the city center is located, being flooded once again. After completing this process, the natural storage areas are now filled, all the catchment is channeled, there are no remaining flood plains, the canals have reached their maximum capacity, and the floodwater is redirected downstream, causing extensive flooding in metropolitan areas. In addition, the city severely restricts the capacity of the drainage system and offers limited opportunities for additional canalization projects.

4. URBAN DRAINAGE DESIGN

Designing urban drainage is a relatively straightforward undertaking, especially when considering the implementation of a project in a new metropolitan area. Channels and pipelines are combined to transport the anticipated flow rate for the planned rainfall. Predictive methods can be used to anticipate infiltration measures and/or reservoirs in order to maintain the generated discharge within a specific threshold. Additionally, additional forms of controls can be implemented to ensure a low effect development. In this scenario, the design is created by merging and totaling sub-catchments in a predetermined sequence. This method calculates the urban land's contribution to the drainage network. When an existing drainage system fails, it might be challenging to provide effective corrective strategies for rehabilitating the system without taking into account its systemic behavior. The amalgamation of surface-generated currents with drainage network currents can be varied. Water discharged from the storm drain system may traverse roadways in an unforeseen manner. The streets transport these flows to lower sections of the basin, rejoining the drainage network without regulation. Occasionally, this shallow movement may extend to an adjacent drainage area, gaining access to other pipes that have not yet been submerged. At times, water can be temporarily accumulated in lower open spaces such as parks or squares, as well as inside structures, in an undesirable manner. Urban flood control involves the rearrangement of flow patterns in both space and time (Canholi, 2005). Assessing the combination of effects in a watershed at a large scale is a challenging task, and there are instances where the offered solutions may not yield the desired results. Mathematical modelling can be a valuable tool in aiding the development of integrated urban flood control solutions.

4.1. Urban drainage traditional design

Conventional urban drainage planning methods rely on the construction of canals to accommodate and manage the concentrated flow of water. This method considers the negative outcomes of the flooding process, specifically the increased and accelerated water flow caused by human-made structures.

The urban drainage system consists of two primary subsystems: micro-drainage and macro-drainage. The micro-drainage system is primarily determined by the arrangement of roadways in urban areas, serving to gather rainwater from urban surfaces. The macro-drainage system is designed to collect and carry away the surface runoff that is brought by the micro-drainage network. Macro-drainage refers to the primary network of drainage, which includes rivers and additional infrastructure such as man-made canals, storm drains, dikes, and other manmade structures.

In general terms, the urban drainage system design comprises the following steps: subdivision of the area into sub-catchments; design of the network integrating urban patterns and natural flows; definition of the design rainfall, considering a certain time of recurrence and a critical time of duration, associated with the concentration time of each sub-catchment considered; step by step calculation of design discharges for each drainage network reach through the Rational Method or another convenient hydrological method; hydraulic design of each drainage network reach. Figure 2 illustrates this approach.

This approach greatly simplifies the real situation and focus only in conveying discharges. Although it may be useful in certain design situations, spatial and temporal effects combination are main factors to be considered when urban floods occur. It is important to have the assistance of a mathematical model as an assessment tool and solutions should be addressed to the catchment as a whole.

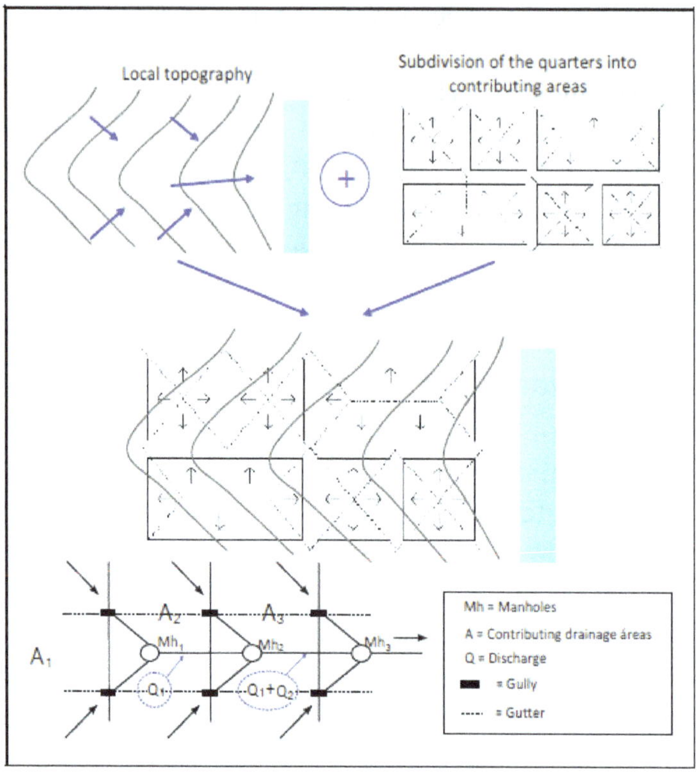

Fig. 2. Schematic urban drainage system classic design.

4.2 Urban drainage design trends

The conventional method of designing drainage systems is being augmented or substituted by more recent concepts that aim to find comprehensive solutions, involving decentralized measures across the entire catchment area, in an effort to restore flow patterns resembling those that existed before urbanization. Storage and infiltration measures are combined in integrated layout solutions. Furthermore, these emerging patterns raise issues with the regulation of water quality, while also promoting the utilization of rainfall as a valuable asset in a comprehensive strategy for the sustainable administration of urban stormwaters. In addition, the potential to integrate flood control measures with urban landscape interventions, which can enhance urban spaces and serve multiple purposes, is increasingly seen as an appealing option for revitalizing deteriorated areas and maximizing the utilization of public investments.

The vision of integrating urban drainage projects with urban development plans and land use and occupation management, provides a better temporal and spatial range of action for flood control projects, as it seeks to intervene not on the consequence of heavy rains, but on the inundation causes. The changing to a point of view of more sustainable solutions on urban drainage requires a commitment with the future consequences concerning the decisions taken today; so solutions must be flexible enough to allow possible modifications and adaptations in the course of urban development (Canholi, 2005).

In urban drainage, sustainability implies that urban floods may not be transferred in space or time. Urban drainage systems have to be planned in an integrated way with urban growth and drainage solutions should be integrated with urban landscape (Miguez et al., 2007). In this context, urbanisation process and urban land use control have both to be thought in order to minimise impacts over the natural hydrological cycle.

This discussion leads to an important point: the understanding on how urbanisation interferes with flow patterns is necessary to develop strategies for stormwater management and urban floods control, by one side, and to establish urban development standards on the other side. Urban drainage planning must consider a broad set of aspects and has to be integrated with land use policy, city planning, building code and all the related legislation. It is possible to say that urban flood control demands the adoption of a varied set of different measures and concepts. Among these measures it is possible to distinguish two greater groups of possible interventions: the *structural measures* and the *non-structural measures*.

Structural measures introduce physical modifications on the drainage net and over urban catchment landscapes, like canalisation, dams, reservoirs, urban flood parks, dykes, among others, intending to change the relations between rainfall and runoff and to reorganise flow patterns. *Non-structural measures* work with environmental education, flood mapping, urbanisation and drainage planning for lower development impacts, warning systems, flood proofing constructions, and other actions intending to allow a more harmonic coexistence with floods.

Structural measures are fundamental when flood problems are installed, in order to bring the situation back to a controlled one. *Non-structural measures* are always important, but are of greater relevance when planning future scenarios, in order to obtain better results, with minor costs.

From this new standpoint, urban drainage projects should aim to counteract the impacts of urbanization and restore the hydrological conditions that existed before urbanization. This will improve the quality of life for the population and promote environmental preservation.

4.2.1 Structural measures

Structural measures can be categorized based on their effectiveness within the catchment. Tucci (1995) categorizes the measures into scattered measures, measures in the micro drainage, and measures in the macro drainage, as explained below and illustrated in Figure 3.

- *Distributed Measures*: these measures act on the lot, squares and sidewalks. They are also known as source control measures.
- *Measures in the micro drainage*: these measures act on the resulting hydrograph from one or more lots.
- *Measures in the macro drainage*: these measures act on the rivers and channels.

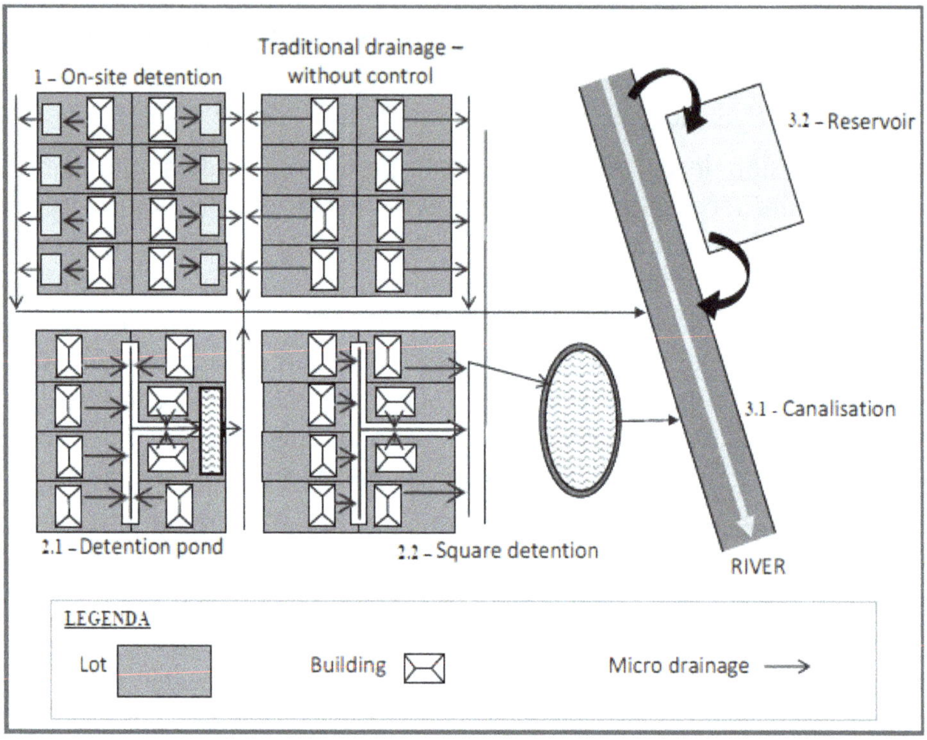

Fig. 3. Examples of flood control measures according to their working principle: 1. Distributed Measures; 2. Measures in the micro drainage; 3. Measures in the macro drainage (Rezende, 2010).

Undoubtedly, canalisation is the more classic method used in flood control operations. The primary goal is to enhance the hydraulic discharge capacity of the macro drainage network by eliminating flow blockages in the main channels, rectifying the river channel, and reinforcing the riverbanks. Another conventional method commonly employed to manage river overflow is the construction of dykes in conjunction with polders, particularly in low-lying regions of the watershed, which facilitate the safeguarding of urbanized lowlands. The protected zones, which cannot independently remove the accumulated water from local catchments during river floods, are typically connected to the main water body through unidirectional gates (FLAP gates) or pumping stations. Therefore, it is imperative to conserve vacant spaces within the polder in order to accept and briefly retain these fluids.

Another set of actions, in addition to enhancing transportation, suggests utilizing storage and infiltration capabilities. Detention ponds are significant examples of this collection of measures. These efforts were typically implemented in the less densely populated areas in the beginning of urbanized zones, when there is more open space to create ponds. In densely urbanized areas where all available space is occupied, detention ponds have been modified to accommodate different sizes. This allows for the utilization of public spaces like parks, parking lots, and squares to temporarily store rainwater from infrequent events. Additionally, this adaptation enhances the urban environment and region, as demonstrated in Figure 4. The image depicts a detention pond in Santiago, Chile, which is integrated with a landscaping design. This type of construction has a broad range of applications and can be utilized by implementing big ponds or by distributing several devices over the watershed. These devices can function inside squares or even within individual lots.

Fig. 4. A detention pond in Santiago, Chile (Photo Miguez, 2009).

Another potential option for stormwater storage techniques could involve the utilization of reservoirs with the objective of enhancing water quality. The buildings in question are the retention ponds and the built wetlands. The primary goal of these procedures is to treat rainwater and minimize its quantitative impact. This is mostly caused by the necessity to establish a permanent reservoir and increase the duration of water retention within the reservoirs in order to facilitate the treatment procedures with the necessary effectiveness. Figure 5 depicts an image of a retention pond that was built in the city of Lagord, France. This pond is an integral component of the comprehensive drainage system for the entire La Rochelle region, encompassing the city of Lagord. Its primary purpose is to facilitate the treatment of rainfall.

Fig. 5. A retention pond in Lagord, France (Photo Rezende, 2009).

Implementing strategies that promote the penetration of rainwater into the earth, so partially restoring the natural hydrology of the watershed, are potentially viable alternatives for flood control. They can adopt various configurations depending on their mode of operation.

A crucial action in this context, particularly due to its environmental consequences, could be the restoration of deteriorated regions, such as hillsides and riverbanks that have been unlawfully seized. According to Miguez & Magalhães (2010), reforestation helps to prevent soil erosion, protect the top layer of soil, and increase infiltration. As a result, there is a decrease in the amount of runoff, which allows drainage structures to function properly by reducing the amount of water and sediments that enter the system.

4.2.2 Non-structural measures

Structural modifications aimed at flood management do not offer comprehensive risk protection for the designated locations. These regions may still experience floods with a magnitude exceeding the amount of protection planned for them. Therefore, it is imperative to implement measures that not only mitigate the threats posed to the community but also assist them in effectively managing flooding. Contrary to structural interventions that physically alter flow relations, non-structural methods aim to minimize the vulnerability of lives and properties to flooding. This set of measures encompasses a diverse array of acts, ranging from urban planning and zoning for floods to the construction of flood-proof structures at an individual level.

Johnson (1978) identified the following non-structural measures: installation of temporary or permanent sealing in the opening parts of the buildings, elevation of pre-existing structures, construction of new elevated structures, construction of small walls or dykes surrounding the structure, relocation or protection of goods that could be damaged within the existing structure, relocation of structures out of the flooding area, use of water resistant material in new structures, regulation of the occupation of the flooding areas, control of new community settlements, regulation of parcelling and building codes, purchase of flood hazard areas, flood insurance, implementation of forecasting and flood warning systems with evacuation plan, adoption of tax incentives for a prudent use of the flood area, and installation of alerts in the area.

5. CONCEPTS OF SUSTAINABLE URBAN DRAINAGE SYSTEMS

The urban drainage system needs to be viewed in an integrated way in the context of the sustainable urban development. It is crucial to understand the crossed relationships between urban growing and flood problems. The aspects involved vary from environmental conservation, land use control, low impact development, and healthy city life. To achieve these goals, related to a sustainable urban drainage, however, it is necessary to construct a framework integrating legal, institutional, social, technical and economical aspects. In this context, it is important to clearly identify the applied regulation in terms of urban zoning and land use control, the water resources policies and the water resources management practices, the integrated environmental sanitation opportunities and constraints, the building standards and limitations, the role of the institutional agents and community participation. Areas to be protected from urban growth need to be delimited, as well as, sometimes, it will be important to recover areas already occupied. This is not simple, due to social pressures against possible dwelling relocation procedures. In developing countries, for example, it is common to have "informal" cities, conforming slum areas growing on risky situation, along riverine areas or hill slopes. Landscape changing should be minimised and original river characteristics could be recovered. In this context, preventing urban land to be heavily impervious is one of the major goals. Minimising impacts on the urban water cycle is of fundamental importance. The development project must comply with natural hydrological aspects or provide compensatory measures for urbanisation changes. The watershed, as a complex and integrated system has to be considered as a whole, not only in physical terms. This must be the unit of planning and design. The documents that integrate urban development and the strategies for a sustainable drainage system, and, therefore, a sustainable city, are the Urban Development Master Plan and The Urban Stormwaters Master Plan.

The non-existence or non-fulfilment of plans for urban development leads to drainage and flood control projects that are restricted to emergency, reactive, and sporadic actions, defined just after the occurrence of disasters (Pompêo, 2000).

This scenario, usually based on the simple and quick removal of water from highly impervious areas by means of canalisations, has become unsustainable and requires a new vision into the problem of urban flooding. Holz & Tassi (2007) suggested that the current drainage system should dismiss the solution of simple removal, as fast as possible, of the non-infiltrated stormwaters that come from the increasing in soil imperviousness, replacing it with measures aiming to mitigate the impacts of this process by facilitating the infiltration and water retention in order to regenerate the hydrological conditions of pre-urbanisation. They emphasize, however, the importance of combining the use of traditional and unconventional drainage structures in order to optimise the system. The fall of the old paradigm should not be simply substituted by a new one. The traditional techniques have to be adapted to a new use, adding the accumulated knowledge to the sustainable solution. Pompêo (2000) emphasizes the need to think the activities related to mitigation of floods in a preventively way, highlighting the value of planning applied to flood control projects. In this context it is introduced the ecosystem approach, which represents the evolution of the reactive thought of the conventional Drainage Master Plan for a proactive and advanced thought in the form of management of the natural and built environment, considering them as interdependent and integrated components. An ecosystem approach can result in lower costs, since it seeks to reduce the need for costly and complex actions of remediation, emphasizing the orientation and planning decisions on land use changes. This option tends also to reduce costs of maintenance over time, because more natural arrangements tend to work by themselves.

This new vision has been based on the concept of Sustainable Development. It's possible to affirm that these systems are designed to both manage the environmental risks of urban stormwater and contribute, whenever possible, to an improved environment and quality of urban life.

Sustainable drainage projects aim to reduce runoff through rainwater control structures in small units. This way, the runoff control performed on source reduces the need for large structures of mitigation and control on the river channels.

Environmental issues presented today to the cities highlight the failure of technical solutions for urban drainage projects, demanding a new approach which should focus on the problem of urban flooding by incorporating the social dynamics and the multisectoral planning in the search of solutions (Pompêo, 1999).

The urban infrastructure systems are interdependent and the fact of not considering the effects of one system over another, or of a system over the urban environment, can reduce the efficiency of these systems or even turn not viable their operation, as is the case of the relations between Drainage and Flood Control systems and Water and Sanitation, Solid Waste, Land Use and Housing systems. Two trivial examples in peripheral countries are the failure of the drainage system operation by the inefficiency or lack of management of municipal solid waste; or the deterioration of health by the inefficient sanitation conditions which, combined with flood events, provide the proliferation of water related diseases.

Problems associated with urban drainage systems are not exclusively technical, but primarily of institutional nature, such as the lack of cooperation between the different departments responsible for urban management and the communication between the city and its citizens (Stahre, 2005). The lack of cooperation between departments may arise from conflicting interests and priorities.

According to Stahre (2005), the Drainage Department of Malmo, Sweden, sees the approach of the sustainable drainage system as an ideal way to achieve its goals and objectives with a low cost, if compared to the traditional drainage system, whereas the Department of Parks and Recreation sees the sustainable drainage solutions as a good ally for the development and improvement of quality of life in the urban environment by increasing the value of urban parks. As a consequence of this sharing of interests, the costs of structures, deployment and maintenance can be divided between the two departments, and also with others that will share the benefits of implementing this system. Stahre (2005) concluded that the solution of rainstorm problems can no longer be regarded as a simple technical service supported only by the Drainage Department, because these waters now represent an important positive resource to the population, inserted in the urban environment. It's important to emphasize the value of water in the city as a demanded resource, not wasting its potential uses.

The failure to consider the drainage in the urban development plans may result in more expensive solutions for flood control, often noneconomical. The successful implementation of a Sustainable Drainage System depends on the cooperation among the different technical departments responsible for urban planning and the active participation of the population.

In this context, the existence of compatibility among the Urban Development Sanitation, Solid Waste and Urban Drainage Master Plans, aiming the integrated planning of the city should be ensured.

The design of the Urban Stormwater Management currently presents, according to Righetto et al. (2009), the aggregation of a structural and non-structural set of actions and solutions, involving large and small works and planning and management of the urban space.

The Stormwater Management Plan of the City must necessarily meet the principles of Sustainable Management of Urban Stormwater, and should seek the following objectives (Ministério das Cidades, 2004):

- Reduce the damage caused by floods.
- Improve the health of the population and of the urban environment, within the economic, social and environmental principles.

- Plan the urban management mechanisms for the stormwaters and the municipality river network sustainable management.
- Plan the distribution of stormwaters in time and space, based on the trend of evolution of the urban occupation.
- Regulate the occupation of areas at risk of flooding.
- Partially restore the natural hydrological cycle, reducing or mitigating the impacts of urbanisation.
- Format an investment program of short, medium and long term.

The drainage projects proposed by the plan should provide the most cost-effective relation, covering social and economic aspects, as well as being integrated into the guidelines of the local River Basin Committee, if any.

The plan must also contemplate a socio-environmental work, through the development of a project that addresses social mobilization, communication, training of educators / agents in the area of environmental sanitation and other environmental education activities, aiming a social-economic and environmental sustainability, including the community participation in the phases of design, implementation, evaluation and use of the proposed works and services.

The premises to be considered in the creation of the Stormwater Management Plan are (Ministério das Cidades, 2004):

- Interdisciplinary approach in diagnosing and solving the problems of flooding.
- Stormwater Plan is a component of the Urban Master Plan. As drainage is part of the urban infrastructure, it should, therefore, be planned in an integrated way.
- Runoff cannot be intensified by the occupation of the basin.
- The Plan has as its planning unit each watershed of the city.
- The stormwater system should be integrated into the environmental sanitation system, with proposals for the control of solid waste and the reduction of stormwater pollution.
- The Plan shall regulate the territory occupation by controlling the expansion areas and limiting the densification of the occupied areas.
- This regulation should be done for each watershed as a whole.
- Flood control is a permanent process and should not be limited to regulation, legislation and construction of protection works. A plan to monitor and maintain the proposed measures is needed a long time.

It is crucial to timely schedule the implementation of the Stormwater Management Plan, taking into account short-term, medium-term, and long-term actions, in order to guarantee sustainable solutions for drainage. The short-term methods aim to address and alleviate the immediate issues with the macro-drainage network by focusing on removing obstructions, clearing sediment, and maintaining the system's natural features. It is crucial to take steps to establish the necessary database for the implementation of the plan. One of the first tasks should be to develop a mathematical model, which will serve as a systematic evaluation tool. Additionally, as part of the short-term efforts, it is imperative to do a diagnostic assessment for the watershed.

Following the initial stage, it is necessary to develop methods in the short/ medium term to manage runoff at its origin and restore the natural characteristics of the hydrological cycle. These interventions should be systematically planned and applied gradually. Flood maps are necessary for assessing the proposed scenarios and determining how they interact with land use zoning.

Based on the accumulated experiences, it is feasible to create a comprehensive drainage manual that provides practical guidance for current and future projects in the medium to long term. The long-term measures should ensure the continued functionality of the drainage system by implementing regular maintenance and monitoring. Furthermore, the implementation of the recommended solutions requires the inclusion of environmental education programs and active involvement of the community.

6. RIVER REVITALISATION

Self-sustaining river systems are recognized for their significant contributions to the ecological and social well-being of human life (Postel and Richter, 2003 apud Palmer et al., 2005). The revitalization of rivers is necessary in order to address the gradual degradation of river ecosystems on a global scale. The results can be examined not only from an aesthetic perspective and to enhance the environment, but also in terms of the hydrological and ecological functioning of restored river sections, augmenting the quantity and quality of river resources and their potential utilization by riverine populations (González del Tánago & García de Jalón, 2007). In urban settings, the process of River Revitalisation is more intricate due to significant alterations in the riverine areas caused by the construction of buildings and highways. These modifications make it challenging to allocate sufficient space for the restoration of the natural functions of the river bed and its banks. It is important to have a focused discussion on the process of revitalizing rivers in urban areas. A solution that balances the natural landscape and the developed environment must be reached through consensus.

In highly urbanised regions, generally the available areas for interventions are scarce, there are numerous socio-economic problems, which make the revitalisation process very complex, because it involves the need of large riparian areas, in order to find space for the river recover its natural course and flooding areas. However, even if these riparian areas were restored to the original natural condition, the heavy modifications that the catchment suffered over the time would probably lead to floods still happening. Thus, the space that would be required to recover the river course functions today is greater than the natural one. Actions in the basin have to be considered, to decrease the imperviousness and to rescue superficial retentions, with the use of reservoirs.

Urban water courses restoration is a challenge for managers, researchers, experts and citizens. In urban environments, the main focus must be on the restoration of the lateral connectivity with the river banks and its tributaries, the restoration of the river natural flow regime, as well as the increase of the degree of freedom of the river. The combination of flood risk management concepts with River Revitalisation measures can be a solution of efficient applicability in urban rivers, in comparison with traditional and localised drainage solutions (Jormola, 2008). Thus, what is expected is the creation of a self-sustainable natural river system, in order to maintain the flood control function after the flow patterns restoration. In this context, a sustainable approach for drainage system should consider River Revitalisation as one of the tools aligned with this major objective.

However, it's important to note that, even when the adopted measures configure only a partial revitalisation, they are important. In addition to reducing the peak flood, they help in the dissemination of this kind of techniques and provide a new perception about the existence of the river for the involved community. They also allow the river valuing and reintegration as part of urban landscape. Finally, it should be pointed out that any process of revitalisation takes time to be fully developed and find it complete. It is necessary to await the responses of the environment, regarding the "new" conditions to which it was submitted. During this time, complementary actions, resulting from the monitoring of this evolution may and should be developed.

7. THE BRAZILIAN EXPERIENCES IN MANAGING THE COMBINATION OF A SUSTAINABLE URBAN DRAINAGE SYSTEM AND WATER RESOURCES ARE SUPPORTED BY INSTITUTIONAL FEATURES.

The establishment of a comprehensive legislative framework, encompassing many institutional and administrative levels, is likely the initial step in achieving sustainability. One of the initial inquiries pertains to the duties of urbanization.

City development falls under the jurisdiction of the Municipality, whereas water resources management requires consideration of the entire basin and typically involves regional planning. In Brazil, legislation demonstrates a notable emphasis on equipping various institutional levels with the necessary tools to facilitate the development of a sustainable city. Overall, Brazilian cities lack a satisfactory level of quality in terms of the physical environment and city life. To gain a comprehensive understanding of the Brazilian legal framework for urban development, specifically in relation to the implementation of sustainable urban drainage systems, a concise overview is provided in the following lines as an introduction to this subject.

The Federal Urban Land Parcelling Act (Brazil, 1979) establishes the minimum standards for urban developing. This Act considers that a plot is a parcel of urban land provided with the basic infrastructure, meeting the urban restrictions. This basic infrastructure refers to urban drainage, sewerage, water supply, electricity and public roads. The Act also states that it is not allowed to have allotments in flooding areas, in environmental protected areas or in polluted degraded areas. Another Federal Act, known as "City Statute" (Brazil, 2001), establishes detailed rules for public land use order and social interest that regulate the use of urban property in favour of the collectivity, the security and well-being of citizens and environmental balance. The urban policy aims to organise the fulfilment of the social functions of the city and of urban property by the application of a set of general guidelines, from which the following topics are detached for the purposes of the present discussion:

- Guarantee the right to sustainable cities, meaning the right to urban land, housing, environmental sanitation, urban infrastructure, transport and public services, work and leisure for present and future generations;
- Democratic management through people's participation and associations representing various segments of the community in the formulation, implementation and monitoring of plans, programs and projects for urban development;
- Planning the development of cities to prevent and correct the distortions of urban growth and its negative effects on the environment;
- Supply of urban infrastructure and community equipments, transport and public services to serve the interests and needs of the population;
- Ordering and control of land use to avoid pollution, environmental degradation and excessive or inadequate use in relation to urban infrastructure;
- Protection, preservation and restoration of the natural and built environment, and cultural, historical, artistic and landscape heritages.

Several important urban management tools are made available in the context of the City Statute. The Urban Master Plan is considered to be the basic instrument for the urban developing policy.

In the field of water resources management the "Water Act" (Brazil, 1997) defines that the hydrographical basin is the unit of planning and design for water resources purposes. Another important reference is the "Basic Sanitation Act" (Brazil, 2007). This Act establishes the first national guidelines for the federal policy of basic sanitation. An important key element of this Act is the integrated conception of the sanitation services and their relation with the efficient management of water resources. For all purposes this Act considers sanitation as a set of services, infrastructure and operational facilities for drinking water supply; sewerage collection, treatment and disposal; urban solid waste management; urban drainage and storm water management.

As it was said in the beginning of this section, although legislation has merits, there are difficulties in urban development and control. This discussion and the possible ways to go further in urban sustainability, concerning urban waters, are the main points of the next topics.

7.1 The local planning and management hierarchy

In federative countries, the municipality's jurisdiction primarily encompasses the provision of local public services, as well as planning, supervision, and development functions. These functions include land use planning, environmental protection, and a certain degree of regulation of economic activities (Dourojeanni and Jouravlev, 1999). In the Brazilian context, municipalities with greater financial resources have recently started taking on responsibilities linked to social services that were formerly handled by the state and federal governments.

It is observed from the 1990's a tendency to extend the role of local public levels regarding the environment management. A lot of factors, however, limit the performance of the municipality in the water management. In Brazil, for instance, there are legal constraints determined by the Federal Constitution, where the cities cannot directly manage the water resources contained within their territories, except for certain formal agreements that transfer some assignments through cooperation arrangements with the State or the Union. Water resources management is a matter of river basin management and cannot be restrained to an administrative territory jurisdiction. Thus the role of the cities is restricted to lower levels of relevance and administrative autonomy (Jouravlev, 2003). Municipality participation in basin organisms for water resources management (called Basin Committees, in Brazil) has been the main stage where the interactions between Municipalities and the other public or private actors occur.

Despite of the fact that the administrative level of the municipality is the closest to social reality, its range of political and administrative roles does not allow a systemic vision of the territory in which it is inserted. In turn, the absence of a clear definition of the extent of local governments functions, in general, linked to the traditional tasks of territorial administration, the supervision and provision of local services, and the fact that most of the municipalities have a reduced financial autonomy, depending on transfers from other government levels, makes it difficult to them to have a more effective participation in the water management. Referring to the financial constraints, Lowbeer & Cornejo (2002) warn that the multilateral financing agencies, except for the Global Environment Facility -GEF, have not yet come to explicit in their agenda the need for integrated management projects of natural resources linked to the territory management and land use, particularly in urban areas. Few are the experiences implemented which have coordination between water conservation/preservation and regulation of the land use against the (dis)functions of urban growth.

Another aspect is that the essentially local nature of the city governments' interests makes them act more like water resources users rather than managers of these resources (Jouravlev, 2003). These aspects are exacerbated in metropolitan areas where local governments have often antagonistic interests and priorities, creating dissent environments with little room for cooperation. The metropolitan question is surely one of the greatest present challenges.

Even if there are restrictions to the municipalities' participation as direct managers of water resources, there is no doubt about the importance of local governments on planning and ordering the territory, due to its consequences on the water resources conservation. It is the county attribution the elaboration, approval and supervision of instruments related to land use zoning and planning for development purposes, such as the Urban Master Plan, the delineation of industrial, urban, rural and environmental preservation areas inside the municipality, land parcelling criteria, the development of housing programs, among other activities that have impact on water resources and sanitation conditions, especially in predominantly urban watersheds.

Peixoto (2006) argues that the history of urban space construction and its effect on natural resources and the quality of human dwellings highlights the challenges in integrating urban and environmental concerns. Simultaneously, there is a noticeable tendency of these challenges aligning with sustainability, as evidenced by the federal Acts mentioned earlier in this area. However, what can be seen in the country is a lack of coordination between the practical tools used for water management and land use planning. This may be a result of the planning process in Brazilian towns lacking legitimacy, as well as specific deficiencies in local legislation.

Many cities exhibit a significant degree of informality and even illegality in their land use practices. This leads to social difficulties and a deterioration of the urban fabric, as slums proliferate and various environmental impacts are exacerbated. Tucci (2004) states that the main challenge in implementing integrated planning is the municipalities' insufficient institutional capacity to handle complex and intersectoral matters. It is important to note that cities might vary based on factors such as their size, geographic location, distance from metropolitan centers, historical urban development, and the qualifications of their technical public staff. Peripheral regions in urban areas sometimes have obsolete legislation, which is worsened by the lack of relevant information and inadequate technological support.

7.2 Integrated water resources management: interfaces with sectoral policies and territory planning

The institutionalization of water resources management in Brazil commenced in the 1930s through the enactment of the Water Code in 1934. The Water Code was a significant step in establishing the water planning process in the country, which facilitated the growth of the energy sector. The Union was granted the authority to create hydroelectric developments and provide energy distribution services by the enactment of this Act. In the same year, the National Department of Mineral Production (DNPN) was established under the Ministry of Agriculture. This department absorbed the Service of Geology and Mineralogy as well as the Water Service. throughout 1938, the National Council of Water and Power was established under the Presidency. Along with the DNPN, it was responsible for making decisions on water and electrical energy throughout the country.

Even before the 30's, several governmental committees had been set up in order to coordinate and implement water works. However, the onset of a coordinated action in the water sector has only occurred in 1933, with the creation of the Sanitation Committee of Baixada Fluminense, in the framework of National Department of Ports and Navigation. Baixada Fluminense is an important region, in the metropolitan area of Rio de Janeiro City, characterised by extensive lowlands. This Committee was responsible for the formulation of a comprehensive drainage program for Baixada Fluminense, which was an unprecedented action in the country, with the main aim to make this vast lowland plains of the State of Rio de Janeiro arable and, secondarily, to eradicate yellow fever and to control the local floods (Carneiro, 2003).

This Committee was the seed of the National Department of Sanitation Works - DNOS (in Portuguese), in 1940, created with the responsibility for implementing the national policy of general sanitation, both in rural and urban spaces, including flooding mitigation, erosion and water pollution control, and the recovery of areas for agricultural or industries uses, as well as the settlement of water supply and sewage systems. Despite the range of assignments given to DNOS, its performance was limited, in its initial phase, to the drainage work for drying wetlands, consolidating and expanding the program prepared by the Sanitation Committee of Baixada Fluminense (*Ibid*).

Dourojeanni and Jouravlev (2001), referring to the experiences of integrated management in Latin America, point out that many of the institutions set up from the 1940's were increasingly incorporating multiple uses of water, even though they used to have as initial particular goals - like flood control, hydroelectric plants, irrigation projects and water supply. Few were those which began their activities by integrating these multiple uses of water.

An important experience in water management in the country, considering the perspective of integrated water resources management, was the creation of the Special Committee for Integrated Watershed Studies (CEEIBH, in Portuguese), in 1978. Supported by CEEIBH, several watersheds committees were constituted. However, despite the important role of these committees on the preparation of studies and investment plans for the recovery and management of the related watersheds, the efforts and experiences have not been able to establish an integrated management of the water resources, nor the implementation of proposed actions could reverse the basins degradation. They also failed to avoid sectoral and fragmented management practices. In part, the low effectiveness of these initiatives was due to the fact that these committees had a merely advisory feature.

Until early 1985, the National Department of Water and Electrical Energy (DNAEE, in Portuguese) was responsible for the management of water resources in the country. Since 1988, however, with the new Federal Constitution, a set of modifications in the water sector was introduced: the definition of Federal/State dominion of the water bodies, the definition of the water as a public good endowed with economic value and the need for the integration of water resources management with land use management policies. In 1995 it was created the Water Resources Secretariat, linked to the Ministry of Environment, with the objective to act in the planning and control of the actions related to water resources in the Federal Government, among others. This institutional change represented the incorporation of the concept of multiple uses of water in the environmental context (CEPAL, 1999).

With the approval of Act 9,433, in 1997, the country received one of the most complete regulatory frameworks focused on water resources managing in the international scenario. The National Water Resources Management System aims to coordinate the administration of water resources in the country seeking to integrate it with other sectors of the economy; administratively arbitrate conflicts related to water uses; implement the National Water Resources Policy; plan, regulate and control the use, preservation and restoration of water resources; and charge for water use, among others. This Act establishes that the watershed is the territorial unit for the National Water Resources Policy implementation and for the National Water Resources Management System actions.

The principles adopted by the Water Act, as the Act 9.433 became known, are adherent to the statements of the major international conferences that have dealt with the water issues and which substantially contributed to the concept of development on a sustainable basis. However, as several authors emphasize (Dourojeanni & Jouravelev, 2001; Cepal, 1999), the integrated management of water resources requires a change on planning paradigms, in both public and private levels. Integrating these variables involves working in several areas of public policy, particularly those connected to regional and urban development and the institutional structures that influence such policies. Silva & Porto (2003) identify four types of integration issues that the institutional planning and administration of water resources system encounters:

- Integration among activities directly related to the water use in the basin: water supply, wastewater depuration, flood control, irrigation, industrial use, energy production, in order to optimise the multiple use under the perspective of a joint management of water quality and quantity;
- Regulatory articulation with sectoral systems that do not use directly the water resources, such as housing and urban transportation, in order to prevent excess of imperviousness or urban pollution impacting water sources;
- Territorial integration with the instances of urban planning and management, in order to apply preventive measures in relation to the urbanisation process, avoiding the increasing demand of quantity and/or quality of existing water resources, including flood occurrences;
- Articulation with neighbouring basins, celebrating stable agreements on the current and future conditions of imported and exported flows of the waters used in the basin.

7.3 Emerging organizational structures: hydrological basins and urban regions

The present approach to managing water resources in urban areas assumes a necessary and interconnected planning process for urban development projects. Tucci (2004) suggested an integrated strategy that considers the Urban Master Plan as the core point for addressing several problems such as watershed preservation, sewerage collection and treatment, solid waste collection and disposal, urban drainage, river floods, and land use.

Gouvea (2005) stated that the dynamics of the urban growth, often disordered and even chaotic, was gradually showing the ineffectiveness of many programs and projects implemented in isolated modules and developed from the mistaken idea that the urban reality could be divided and treated in a compartmentalised way. The author notes that actually the city must be seen not only as a specific and complex system, but also as part of a larger system, regional or even national, made up of several subsystems, such as habitation, public transportation, sanitation, natural environment, etc., which are closely related and require an integrated and multidisciplinary approach.

Tucci (2004) lists some factors that hinder the application of the integrated management concepts in cities, as follows:

- Absence of adequate knowledge on the subject: the population and professionals from different fields and levels who do not have adequate information on the problems and their real causes. The decisions usually result in high costs. For example, the use of canalisation works as drainage solutions is a widespread practice in Brazil, even when
- They represent high costs and impacts. Generally, the channels transfer flood downstream, affecting another part of the community. These works can reach an order of magnitude of 10 times the costs of on-source control measures;
- Inadequate design for urban systems control: an important part of the technicians who work in urban issues is outdated about the environmental concerns;
- Fragmented vision of the urban planning: urban planning and development do not always incorporate integrated aspects related to water supply, sewage, solid waste, flooding and urban drainage;
- Lack of management capacity: the cities are not designed for proper management of the different water aspects in urban areas.

The situation is particularly dire in urban regions with a significant degree of conurbation. The emergence of new institutional frameworks for urban administration has attracted the attention of technologists and scholars who see the necessity of reintroducing regional planning, while also acknowledging the importance of addressing local issues. Consequently, the difficulties associated with managing urban waters in conjunction with the rapid expansion of urban areas give rise to distinct issues in the built environment that necessitate a customized strategy. The subsequent sub-items analyze the novel institutional arrangements and the perspectives they offer to address the institutional void resulting from the lack of metropolitan instances for highly urbanized city development in Brazil.

7.3.1 Watershed committees

The watershed committee is the key entity in the water resources management system. The committees are public political entities responsible for making decisions and have legislative, deliberative, and advisory functions. They focus on the use, protection, and restoration of water resources and include a diverse representation of organized sections of civil society, governments, and water consumers.

The committees work as a decentralised locus for discussion on the water uses issues of a watershed, acting as a mediating instance among different interests. These committees are seen as "water parliaments", playing the role of the decision maker within the basin context. The committees composition, as provided by Act 9,433, comprises the Union, the States and the Cities located, even if partially, in the respective basin; the water users within basin area; and the civil water resources entities with activities in the basin.

Nevertheless, it is a fact that the committees established in the country have found great difficulties in fulfilling their decisions and in executing their investment plans. Two main aspects can be identified as constraints to the action of the committees. The first one is that the revenues from charging water uses, which is the only funding source of the committees, are not enough to make the necessary investments for the watersheds recovery. Thus, the committees remain dependent on the traditional sources of investment, which have their own mechanisms for eligibility and prioritisation. The second aspect is that the committees have not gained the necessary political and institutional legitimacy for the public policies coordination related to the watershed, nor could it guide the investments to target actions of its interest. This last aspect stems from the fact that the basin does not constitute a political space reference for Brazilian institutions.

Without disregarding the committees importance in the public policies decentralisation and in the society participation, the above pointed aspects restrict the possibilities of the committees working as integrators of public regional policies.

7.3.2 Public consortia

The formation of consortia in Brazil originated in the late 19th century. However, there have been many configurations and designs of autonomy for these instances of inter-municipal collaboration over time. Table 1 provides a concise overview of the many types of consortia that have been planned in Brazil for more than a century.

Table 1. Consortia Models provided in Brazil in the period 1891-2007.

Period	Organisation Model
1891 - 1937	The consortia were contracts celebrated among municipalities whose effectiveness depended upon state approval.
1937	The Constitution recognises that municipalities association in consortia are legal public entities.
1961	It is created the first Brazilian inter-federative autarchy.
1964 - 1988	Administrative consortia arise as collaboration agreements without legal personality.
From 1998	Creation of several public consortia. The Constitutional Amendment n0 19 changed the art. 24 of the Constitution of 1988, introducing the concepts of public consortia and the associated management of public services.
2005	Public Consortia Act
2007	The Decree 6,017, of 17-01-2007, regulates the Public Consortia Act

Source: Adapted from Rieiro, 2007.

As it can be seen, between 1964 and 1988 administrative consortia arose as simple collaboration agreements, without legal personality, reflecting the period of centralism of the military government. From the 1990s, based on the Constitution of 1988, a great number of public consortia appeared in Brazil, especially in the health field. Consortia were also formed around specific themes, being the most common the regional development and the environment, water resources and sanitation.

Most of the consortia established in the country involve small and medium communities. Only 5% of the consortia include cities with more than 500,000 inhabitants (Spink, 2000, apud Gouvea, 2005). Gouvea (2005) states that the main obstacle to the formation of inter- municipal cooperation is still the autarchic aspect of the Brazilian municipalities, in a 'compartmentalised' federalism context, which rigidly separates the counties. Thus, the Brazilian federative framework does not ease the cooperation between municipalities.

The discussion about the new Public Consortia Act began in August 2003, aiming to regulate the Article 241 of the Constitution and give more legal and administrative security to the partnerships among the consortium parts. In 2005, the Congress approved the new Act. The public consortia, according to this Act, are partnerships formed by two or more entities of the federation to achieve common interest goals, in any area. The consortia can discuss how to promote the regional development, manage solid and wastewaters disposal, and build new hospitals or schools. They have their origin in the municipalities associations, which were already established in the Constitution of 1937. One of the purposes of the public consortia is to enable and make viable the public administration of the metropolitan areas, where urban problems solutions require joint policies and actions.

The collaboration enables small communities to collaborate and, by increasing their size, enhance their technical, operational, and financial capacities. Furthermore, it is feasible to form coalitions in domains of shared concern, such as watersheds or regional development hubs, with the aim of enhancing public services. Undoubtedly, the new Act introduced a promising instrument for addressing common issues in urban regions, providing public organizations with a practical option for collaboration at a higher-than-municipal level.

7.3.3 Regulation of the sanitation sector

After a long period without a regulatory mark for the sanitation services, it was approved, on 5 January 2007, the Basic Sanitation Act, nº 11,445. With this Act, the country established a modern regulatory mark for the sanitation sector, integrated with the National Policy of Water Resources Management, and establishes the national guidelines for the basic sanitation sector.

This Act considers as basic sanitation the public supply of potable water services; the collection, transportation, treatment and adequate sewage final disposal; the collection, transportation, transfer, treatment and final disposal of household waste and garbage originated in the public streets and open areas; the drainage and urban storm water management, considering the conduction, detention or retention of the flood flows and the treatment and final disposal of the rainwater drained from the urban areas.

The Act states in its fundamental principles, among other things, the need of making the sanitation services available for all urban areas, preserving citizens health and public and private structures. The sanitation services need to be articulated with urban and regional policies of development and integrated with water resources management. These principles clearly demonstrate the integrative perspective of the Act, fleeing from the traditional view of the sanitation sector in the country, especially when compared with the actions of the former National Department of Sanitation Works (DNOS).

The Basic Sanitation Act also provides several innovations, among others, the possibility of the holders of public sanitation services delegate these services to Public Consortia.

Another unprecedented advance is the possibility of including in the contracts some progressive and gradual goals for services expansion, increasing in quality, efficiency and rational use of water, and energy rationalisation.

Members of the Federation are also allowed to establish funds, separately or together in public consortia, which may be composed, among other resources, by a portion of the revenue from the services, in order to cover the costs of the universalisation of the public sanitation. These funds, in addition to traditional funding sources, can solve the chronic lack of financing for the sector, especially in relation to urban drainage, where resource allocation is more uncertain.

The Act consolidates the possibility of formation of the Public Consortium for providing regionalised public sanitation services, as stated in the Public Consortium Act.

The Decree nº 6,017 of 17 January 2007, that regulated the Public Consortia Act, details the way public bodies may constitute consortia. The first aspect to be noted is that the public consortium will be constituted as a legal entity formed solely by members of the Federation, organised as a public association, with legal personality under public law and autarchic nature, or as a legal entity of private law with non-profit aims.

The goal of the consortium will be determined by the associated public entities, assuming, among others, the following possibilities:

- Associated management of public services;
- Services provision, including technical assistance, structures construction and goods supply to the direct or indirect administration of the consortium members;
- Sharing or common use of the equipments or instruments, including those for management, maintenance, data processing;
- Technical studies production;
- Water resources rational use promotion and the environment protection;
- Performance of functions in the water resources management system that have been to them delegated or authorised;
- Local and regional policies and actions for urban and socio-economic development.
- The prevision for the exercise of the multi-sectoral functions opens the way for the establishment of a technical agency with legal competence for integrating the public policies involving environment, water resources, sanitation and land use planning within a regional scope.

7.4 Integrated viewpoints: water resources, sanitation, and urban development

Probably, the most urgent and complex task of the agenda of public managers really committed to build a sustainable future for the cities refers to promote the integration of public policies concerning water resources, sanitation and urban land use ordering.

The metropolitan question is an issue of increasing importance, and the built environment worsens with the growing of the cities and the conurbation process.

At the moment, in Brazil, there are available tools for building institutional arrangements that resume the management in metropolitan basis, replacing the model that prevailed in the last twenty years, which focused in local and fragmented policies. Thus, there are reasons to believe that the new institutional arrangements figuring in the country offer alternatives for the shared management between states and municipalities, especially in the larger urban agglomerations. The public consortia may have more political and legal legitimacy to plan the interventions that could cause better impacts in the territory, in an integrated manner, interacting with all the levels of government and society. It is strongly needed the long-term planning resumption, based on effective cooperation mechanisms.

Specifically regarding the action of the municipalities, there is a vast field of possibilities to be sought, especially after the approval of the City Statute. The new Master Plans, previewed in this statute, can and should incorporate mechanisms for a more effective management of land use, using a wider range of legal, economic and tax-oriented instruments for the urban development on a sustainable basis.

Finally, the improvement of the technical management of cities and metropolitan areas challenge remains. Again, it is stressed the need for creation of cooperative structures, not only among the various municipalities of the same metropolitan area, but also between these municipalities and the state, for the definition and implementation of policies in an integrated manner.

8. CASE STUDY: THE SEEK FOR SUSTAINABLE SOLUTIONS FOR URBAN FLOODS IN IGUAÇU- SARAPUÍ RIVER BASIN AT RIO DE JANEIRO STATE, BRAZIL

A case study regarding the Iguaçu-Sarapuí River Basin, located in the western portion of the Guanabara Bay Basin, which lies in the Metropolitan Region of Rio de Janeiro is discussed. Figure 6 shows a map of this area, with the Cities that are in this basin. This is one of the most critical areas in the state regarding urban flooding. This region is densely occupied and presents great urban and industrial development areas, as well as wide rural zones in an urbanising process, and reminiscent areas of natural vegetation on the upstream reaches of the basin. This case study intends to illustrate and complement the conceptual discussion held in this chapter and to show how complex the interaction of urban drainage problems and the city growth can be in a context of an unplanned and non integrated reality. In this region, urban expansion dynamics is, in general, marked by irregular occupation, in terms of land tenure and urban regularisation, and lack of sanitation.

The region under consideration has a great portion of Baixada Fluminense Lowlands. Interventions for flood mitigation were supported by Federal Government on 1930's (canals construction, dams, floodgates and pumping stations). At that time, the hydraulic structures were projected for agricultural uses. A migratory process for this area began on the 1950 's and accelerated from the 1970's on. In the beginning of the 1990 decade, Baixada Fluminense Lowlands sheltered more than 2 million inhabitants in 6 counties. More than 350 thousand of these inhabitants suffered the effects of significant floods. The chaotic process of urbanisation resulted in the occupation of the main rivers bed, what has made almost impossible the maintenance of the watercourses; the acceleration of the process of rivers and canals sedimentation due to deforestation of the slopes and inadequate solid waste disposal; and the increase of the runoff, due to uncontrolled vegetal removal and consequent substitution by impervious surfaces.

Fig. 6. Iguaçu-Sarapuí River Basin at Baixada Fluminense Lowlands in the Metropolitan Area of Rio de Janeiro.

The basin of Iguaçu-Sarapuí River, in the past years, became a stage for articulated actions focusing on flood control and environmental recovery, in the context of the revision of its Water Resources Master Plan, in a study made by the Federal University of Rio de Janeiro for the State Institute of the Environment (INEA, in Portuguese). This study was supported by a mathematical model, called MODCEL, capable to represent the system in integrated terms (Mascarenhas & Miguez, 2002). Figure 7 shows the mapped flood conditions for present situation, represented over the modelled area of Iguaçu River Basin.

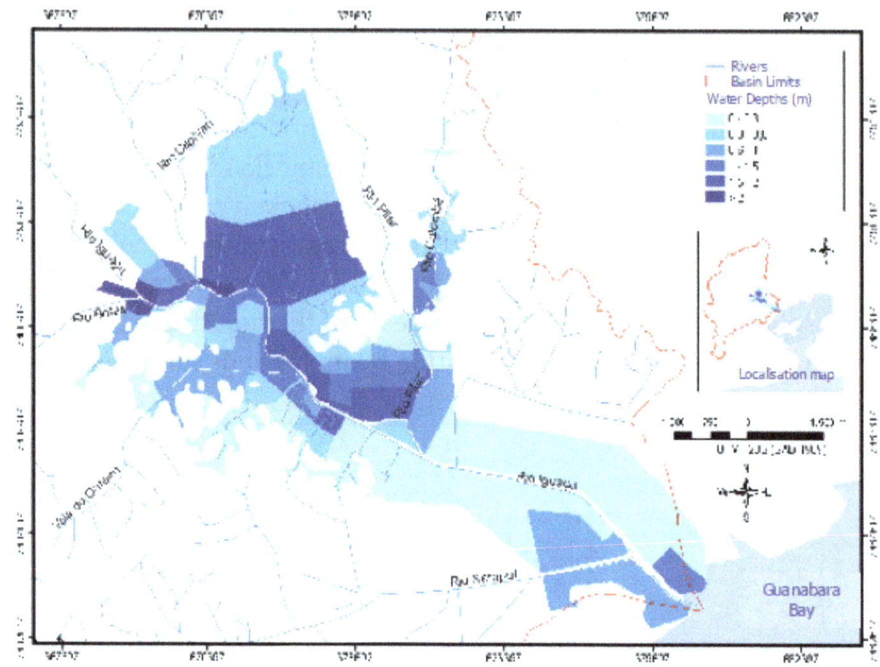

Fig. 7. Flood map of Iguaçu River Basin, for a design rainfall of 20 years of return period, calculated with a mathematical model aid.

Some of the proposed actions were:

- Maintenance of spaces free of urbanisation, preventing the aggravation of flooding at the consolidated urban areas;
- Land use regulation and control, by means of the establishment of Environmental Preservation Areas;
- Implementation of urban parks;
- Creation of public consortiums for integrated planning of policies for multi-counties interests (recognising the importance of the metropolitan planning);
- Revision and adaptation of the municipalities urban planning instruments. Three kinds of parks were designed:

1. **Fluvial Urban Park** - longitudinal parks along rivers, with the purpose of protecting water course banks, also avoiding their irregular occupation by low income population.
2. **Flooding Urban Park** - longitudinal parks implemented in low elevation areas to allow frequent inundations, which will help to damp flood peaks.
3. **Environmental Urban Park** - parks with greater dimensions, flat or not, with the purpose of environmental preservation and land use valuing, aiming to minimise runoff generation and maintaining a buffer of pervious surfaces.

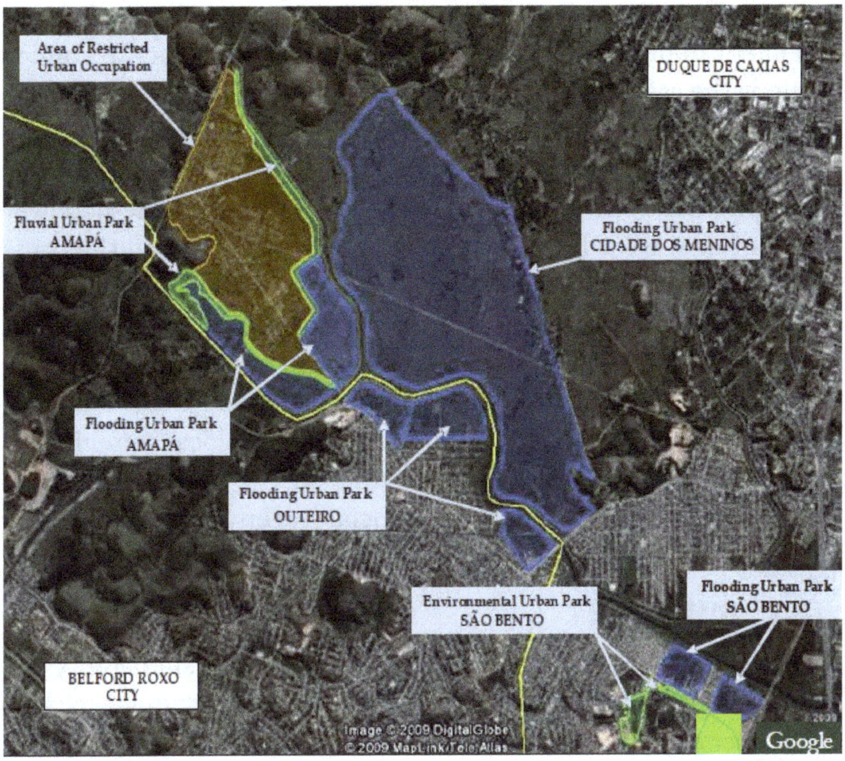

Fig. 8. Shows the general propositions for Iguaçu River Basin, while Figure 9 shows two more detailed examples of the proposed parks.

The State carries out complementary actions such as collaborating with each Municipality in the basin to implement the proposed measures, establish local regulations for urban land use, and conduct environmental education campaigns. These initiatives are funded by the Federal Government through a specific Program of Developing Acceleration.

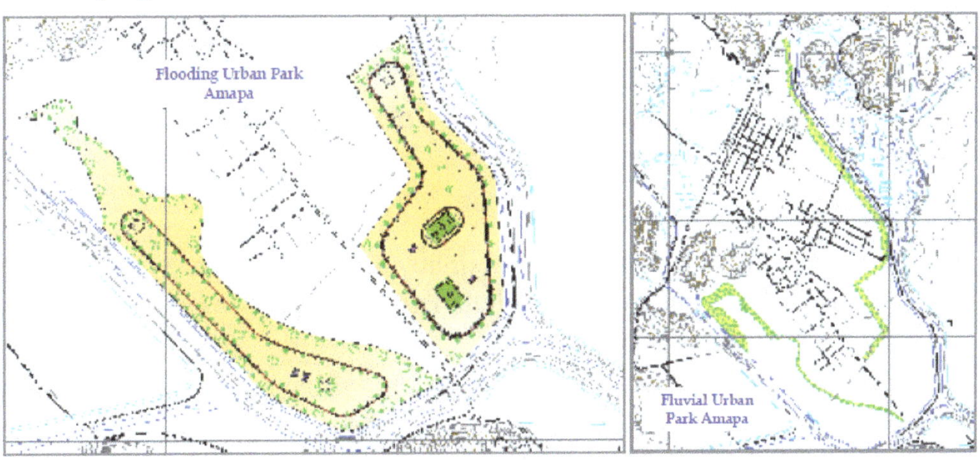

Fig. 9. Fluvial Urban Park and Flooding Urban Park examples.

Based on information from INEA, these interventions will directly and indirectly benefit approximately 3 million residents of Baixada Fluminense. The benefits include the restoration of 80 km of deteriorated riverbanks, the relocation of 2,200 families from high-risk areas to new housing developments in nearby regions, the establishment of parks and recreational areas to safeguard the restored riverbanks from future encroachments and to temporarily store floodwaters, the identification of new areas for environmental preservation, and the construction or restoration of approximately 70 km of roads within the intervention areas, including the rehabilitation of narrow crossings, bridges, aqueducts, gates, and polders. Furthermore, it is projected that over 200,000 trees would be planted along the riverbanks. INEA has been implementing a monitoring and evaluation process for this project by engaging local communities through local committees and the regional forum of participation and social control. The aim is to raise awareness and mobilize these communities regarding social and environmental issues, as well as emphasize the significance of participation and social control.

9. CONCISE SUMMARY

Urban expansion is a global phenomenon, and ensuring sustainability is a crucial aspect that must be prioritized in the future. Generally, huge cities provide numerous challenges such as the need for land use management, substandard living conditions, high unemployment rates, poverty, inadequate transit systems, insufficient public services, and a lack of infrastructure, among other issues. The issue of water resources management and sanitation is crucial in this context. The issue of urban flooding is undeniably one of the foremost issues that cities will inevitably confront. Urbanization is a significant anthropogenic factor that greatly impacts the occurrence of floods. Conversely, inside an urban setting, the occurrence of floods is one of the primary factors that causes significant deterioration. When examining urban drainage within the framework of integrated city development, the sustainability perspective presents a wide range of possibilities for integrated solutions. These solutions can be explored in various areas such as hydraulic engineering, architecture and urbanism, city planning and management, social disciplines, and economic considerations.

REFERENCES

Andoh, R. Y. G. & Iwugo, K. O. (2002). Sustainable Urban Drainage Systems: - A UK Perspective. *Proceedings of the 9th International Conference on Urban Drainage*, Portland, Oregon, USA, 2002.

Arzet K. (2010). O Rio Isar; Munique, Alemanha, In: *Revitalização de Rios no Mundo: América, Europa e Ásia.* Machado A.T.G.M., Lisboa A.H., Alves C.B.M., Lopes D.A., Goulart E.M.A., Lite F.A., Polignano M.V., pp 153-168, Instituto Guaicuy, ISBN 978.85.98659.08.4, Belo Horizonte, Brazil. (in Portuguese)

Batista, M.; Nascimento, N. & Barraud, S. (2005). *Técnicas Compensatórias em Drenagem Urbana*, ABRH, ISBN 858868615-5, Porto Alegre, Brazil.

Benevolo, L. (2001). *História da Cidade,* Editora Perspectiva, São Paulo, Brazil.

Biswas, A.K. (1970). *History of Hydrology*, North Holland Publishing Company, Amsterdam. Brazil. (1979). *Federal Act 6,766, of December 19, 1979.* Federal Official Gazette of Brazil, Brasília, DF, 20 December. 1979. Section 1, Brasília, Brazil. (in Portuguese)

Brazil. (1997) *Federal Act 9,433, of January 8, 1997.* Federal Official Gazette of Brazil, Brasília, DF, n. 6, 09 January. 1997. Section 1, Brasília, Brazil. (in Portuguese)

Brazil. (2001) *Federal Act 10,257, of July 10, 2001.* Federal Official Gazette of Brazil, Brasília, DF, n. 133, 11 July. 2001. Section 1, Brasília, Brazil. (in Portuguese)

Brazil. (2007) *Federal Act 11,445, of January 5, 2007.* Federal Official Gazette of Brazil, Brasília, DF, n. 8, 11 January. 2007. Section 1, Brasília, Brazil. (in Portuguese)

Burian, S.J. & Edwards, F.G. (2002). Historical perspectives of urban drainage, Global Solutions for Urban Drainage, *Proceedings of the 9th International Conference on Urban Drainage*, Portland, September 2002.

Canholi, A.P. (2005). *Drenagem urbana e controle de enchentes*, Oficina de Textos, ISBN 8586238430, São Paulo, Brazil.

Carneiro, P. R. F. (2003). *Dos Pântanos à Escassez: uso da água e conflito na Baixada dos Goytacazes*, Annablume, São Paulo, Brazil.

CEPAL – Comisión Económica para América Latina y el Caribe. (1999). *Tendencias actuales de la gestión del agua en América Latina y el Caribe (avances en la implementación de las recomendaciones contenidas en el capítulo 18 del Programa 21.* LC/L.1180, Santiago de Chile.

Chocat, B., Krebs, P., Marsalek, J., Rauch, W. & Schilling W. (2001). Urban Drainage Redefined: from Stormwater Removal to Integrated Management. *Water Science and Technology*, Vol. 43, No. 5, (2001), pp. (61–68).

CIRIA (2007). *The SUDS Manual*, by Woods-Ballard, B.; Kellagher, R.; Martin, P.; Bray, R.; Shaffer, P. CIRIA C697.

Coffman, L.S., Cheng, M., Weinstein, N. & Clar, M. (1998). Low-Impact Development Hydrologic Analysis and Design. *Proceedings of the 25th Annual Conference on Water Resources Planning and Management*, Nova York, USA, 1998.

Dourojeanni, A. & Jouravlev, A. (1999). *Gestión de cuencas y ríos vinculados con centros urbanos.*

CEPAL - Comisión Económica para América Latina y el Caribe, 1999.

Dourojeanni, A. & Jouravlev, A. (2001). *Crisis de Gubernabilidad en la Gestión del Agua. Serie Recursos Naturales e Infraestructura,* No. 35, Cepal, División de Recursos Naturales e Infraestructura, Santiago.

González del Tánago, M. & García de Jalón, D. (2007). *Restauración de ríos. Guía metodológica para la elaboración de proyectos.* Ministerio de Medio Ambiente, Madrid, España. (in Spanish)

Gouvêa, R. G. (2005). *A questão metropolitana no Brasil*, Editora FGV, Rio de Janeiro.

Gusmaroli, G.; Bizzi, S. & Lafratta, R. (2011) L'approccio della Riqualificazione Fluviale in Ambito Urbano: Esperienze e Opportunittà. *Anais do 4° Convegno Nazionale di Idraulica Urbana*, Veneza, Itália, June 2011.

Hill, R. (2010) O Rio Tâmisa: Londres, Inglaterra. Machado A.T.G.M., Lisboa A.H., Alves C.B.M., Lopes D.A., Goulart E.M.A., Lite F.A. and Polignano M.V. (Org.) In: *Revitalização de Rios no Mundo: América, Europa e Ásia.* Machado A.T.G.M., Lisboa A.H., Alves C.B.M., Lopes D.A., Goulart E.M.A., Lite F.A., Polignano M.V., pp 131- 152, Instituto Guaicuy, ISBN 978.85.98659.08.4, Belo Horizonte, Brazil. (in Portuguese)

Holz, J. & Tassi, R. (2007). Usando Estruturas de Drenagem Não Convencionais em Grande Áreas: O Caso do Loteamento Monte Bello. *Proceedings of the XVII Simpósio Brasileiro de Recursos Hídricos*, São Paulo, Brazil, November 2007.

Johnson, W.K. (1978). *Physical and economic feasibility of nonstructural flood plain management measures*, Institute for Water Resources, U. S. Army Corps of Engineers, Fort Belvoir, VA.

Jormola, J. (2008). Urban Rivers, In: *Proceedings of the 4th ECRR Conference on River Restoration*, Venice S. Servolo Island, Italy, June 2008.

Jouravlev, A. (2003). Los *municipios y la gestión de los recursos hídricos.* Serie Recursos Naturales e Infraestructura. CEPAL - Comisión Económica para América Latina y el Caribe, nº 66, 2003.

Langenbach, H.; Eckart, J. & Schröder, G. (2008). Water Sensitive Urban Design – Results and Principles. *Proceedings of the 3rd SWITCH Scientific Meeting*, Belo Horizonte, Brazil, 2008.

Lowbeer, J. D., Cornejo, I. K. (2002). *Instrumentos de gestão integrada da água em áreas urbanas.* Subsídio ao Programa Nacional de Despoluição de Bacias Hidrográficas e estudo exploratório de um programa de apoio à gestão integrada. USP/ Núcleo de Pesquisa em Informações Urbanas - Convênio FINEP CT-HIDRO.

Mascarenhas, F.C.B. & Miguez, M.G. (2002). Urban Flood Control Through a Mathematical Cell Model. *Water International*, Vol. 27, Nº 2, (June 2002), pp. (208-218).

Miguez, M.G., Mascarenhas, F.C.B., Magalhães, L.P.C. (2007). Multifunctional Landscapes For Urban Flood Control In Developing Countries. *International Journal of Sustainable Development and Planning*, Vol. 2, Nº2 (2007), pp. 153-166, ISSN 1743-7601.

Miguez, M.G., Magalhães, L.P.C. (2010). Urban Flood Control, Simulation and Management: an Integrated Approach. In: *Methods and Techniques in Urban Engineering*, Armando Carlos de Pina Filho & Aloísio Carlos de Pina, pp. (131-160), In-Tech, ISBN 978-953- 307-096-4, Vukovar, Croatia.

Ministério das Cidades. (2004). *Manual de Drenagem Urbana Sustentável*, Brasília, Brazil, 2004.

Noh, S.H. (2010). Rio Cheonggyecheon: Seul, Coreia do Sul. Machado A.T.G.M., Lisboa A.H., Alves C.B.M., Lopes D.A., Goulart E.M.A., Lite F.A. and Polignano M.V. (Org.) In: *Revitalização de Rios no Mundo: América, Europa e Ásia*. Machado A.T.G.M., Lisboa A.H., Alves C.B.M., Lopes D.A., Goulart E.M.A., Lite F.A., Polignano M.V., pp 291-304, Instituto Guaicuy, ISBN 978.85.98659.08.4, Belo Horizonte, Brazil. (in Portuguese)

Palmer, M.A., Bernhardt, E.S., Allan, J.D., Lake, P.S., Alexander, G., Brooks, S., Carr J.,Clayton, S., Dahm, C.N., Follstad Shah, J., Galat, D.L., Loss, S.G., Goodwin, P., Hart, D.D., Hassett, B., Jenkinson, R., Kondolf, G.M., Lave, R., Meyer, J.L., O'Donnel, T.K., Pagano, L. & Sudduth, E. (2005). Standards for ecologically successful river restoration. *Journal of Applied Ecology*, Vol. 42, Nº 2, (April 2005), pp. 208-217.

Peixoto, M. C. D. (2006). Expansão urbana e proteção ambiental: um estudo a partir do caso de Nova Lima/MG. In: *Novas Periferias Metropolitanas - A expansão metropolitana em Belo Horizonte: dinâmica e especificidades no Eixo Sul*, Heloisa Soares de Moura Costa (organizadora); Geraldo Magela Costa, Jupira Gomes de Mendonça, Roberto Luis de Monte-Mór (colaboradores); [Editor: Fernando Pedro da Silva] - Belo Horizonte: C/Arte, 2006.

Pompêo, C. A. (1999). Development of a state policy for sustainable urban drainage. *Urban Water*, Vol. 1, Nº 2, (July 1999), pp. 155-160.

Pompêo, C.A. (2000). Drenagem Urbana Sustentável. *Revista Brasileira de Recursos Hídricos*, Vol. 5, Nº 1, (January-March 2000), pp. 15-23.

Rezende, O. M. (2010). *Avaliação de medidas de controle de inundações em um plano de manejo sustentável de águas pluviais aplicado à Baixada Fluminense*. MSc. Thesis, Federal University of Rio de Janeiro, Brazil, 2010.

Righetto, A.M., Moreira. L.F.F. & Sales, T.E.A. (2009). Manejo de Águas Pluviais Urbanas. In: *Manejo de Águas Pluviais Urbanas, Projeto PROSAB*, Righeto, A.M., pp. (19-73), ABES, Natal, Brazil.

Rieiro, W. A. (2007). *Cooperação Federativa e a Lei de Consórcios Públicos*. Brasília - DF: CNM.

Silva, R. T. & Porto, M. F. A. (2003). Gestão urbana e gestão das águas: caminhos da integração. *Estudos Avançados,* Vol. 47, No. 17, (January-April 2003), pp.(129-145).

Stahre, P. (2005). 15 Years Experiences of Sustainable Urban Storm Drainage in the City of Malmo, Sweden, *Proceedings of World Water and Environmental Resources Congress*, Alaska, May 2005.

Tucci, C.E.M. (1995). Inundações Urbanas, In: *Drenagem Urbana*, Tucci, C.; Porto, R.; Barros, M., pp. 15-36, Editora da Universidade/ABRH, ISBN 8570253648, Porto Alegre, Brazil.

Tucci, C. E. M. (2004). Gerenciamento integrado das inundações urbanas no Brasil. *Rega/Global Water Partnership South América*, Vol. 1, No. 1, (January-June, 2004), pp. (59-73).

United States Environmental Protection Agency - US EPA. (2004). *The Use of Best Management Practices (BMPs) in Urban Watersheds,* by Muthukrishnan, S.; Madge, B.; Selvakumar, A.; Field, R.; Sullivan , D. EPA/600/R-04/184S.

Wong, T.H.F. (2006). Water Sensitive Urban Design - the Journey Thus Far. *Australian Journal of Water Resources*, Vol. 10, No. 3, (2006), pp. 213-222.

03 SUSTAINABLE URBAN DRAINAGE SYSTEMS (SUDS)

1. INTRODUCTION: AN OVERVIEW

Water is a naturally occurring resource that is replenished through the physical processes of the hydrologic or water cycle. Solar radiation causes the evaporation of water from the surfaces of oceans, lakes, rivers, and land, resulting in the water transforming into water vapor and returning to the atmosphere. Water is also absorbed by plants, which utilize it to fulfill their physiological requirements and release it through transpiration. Evapotranspiration refers to the complete process of transferring water vapor from the Earth's surface to the atmosphere. Once it is in the atmosphere, it can condense and fall as rain, snow, dew, or frost. Upon reaching a surface, the water can either runoff or permeate through layers of soil. As a result of the specific characteristics of the land's surface, water that flows over the ground comes together in valleys, forming rivers and lakes. These bodies of water then continue to flow into larger bodies of water until they eventually reach the ocean. The infiltrating water has the potential to move downwards into lower soil layers, where it may resurface as springs, or seep farther into subsurface aquifers. An aquifer that is directly connected to the surface is referred to as non-confined. The water in this aquifer is held in a region known as the water table, which is influenced by air pressure. Aquifers are considered confined when there is a geological feature that acts as a barrier between the water storage zone and the soil surface. In these cases, the aquifer experiences a pressure that is higher than atmospheric pressure. The water contained inside either of these aquifers can be released as base flow, which occurs as a result of the topographic gradient, supplying rivers, lakes, and other bodies of water. Indeed, this base flow is accountable for the regulation of river flow during seasons of low precipitation.

The aforementioned mechanism remains in constant operation, driven by solar energy, hence its designation as the hydrologic cycle. However, in order for this cycle to proceed naturally, there must be no changes in the amount of water that remains in each step of the process, namely, in the atmosphere, on the surface, and in the soil.

Indeed, numerous human and non-human activities have the potential to disrupt the hydrologic cycle. Urbanization is one of the most influential human activities that affect the natural hydrologic cycle.

Urbanization results in the soil becoming impermeable due to the process of paving, as well as soil compaction caused by automobiles and buildings. These factors hinder the infiltration of precipitation (Silveira, 2002). The water that used to be absorbed by the soil is now running off surfaces, resulting in increased runoff. This runoff accumulates in lower-lying places, leading to the formation of flood-prone zones. The augmentation of surface runoff can also result in the inundation of streams, gullies, ditches, and rivers, causing issues with riverside flooding.

Flooding poses hazards to individuals' well-being and overall standard of living, as well as causing adverse social and economic consequences. This is a recurring event that takes place almost every year in major urban areas of Brazil. Experts assert that there is a prevailing tendency for the frequency of critical incidents to rise, as presently, nearly every medium-sized city in Brazil has some level of flooding at some point.

Prior to the 1990s, the issue of flooding was addressed using a health services strategy (Silveira, 2002). Previously, flooding issues were resolved by constructing a rainwater drainage system with the aim of enhancing the efficiency of water flow and redirecting rainwater to a different body of water downstream.

Nevertheless, this technique proved inadequate in addressing flooding issues as urban expansion and the subsequent increase in soil impermeability led to the emergence of additional flood-prone areas, necessitating the implementation of increasingly larger water management systems. To clarify, the original hydrologic cycle was completely changed, and efforts were made to improve the efficiency of water flow through this system, despite the escalating expenses involved.

Furthermore, precipitation in urban regions can be tainted while still in the atmosphere if the atmosphere is polluted or can become polluted when it comes into contact with urban surfaces. This contamination occurs as a result of the washing away of substances such as oil, grease, and fecal matter, along with other pollutants. Consequently, the land becomes contaminated and bodies of water become deteriorated, which compromises the availability of surface water.

Amidst the unsustainable scenario throughout the 1990s, several countries, inspired by the United States, France, and Australia, suggested a range of new approaches to effectively manage and purify rainwater in metropolitan areas, both in terms of quality and quantity. The primary concept conveyed in the suggestions put out for rainwater management in urban areas is the preservation of natural water flow mechanisms, or the implementation of structures that aim to replicate a certain component of the natural hydrologic cycle that has been disrupted.

Therefore, the primary methods employed involve the utilization of structures designed to restore the soil's water infiltration capacity that has been diminished as a result of impermeabilization. Consequently, the generation of surface runoff is reduced, leading to a decrease in flooding issues. Moreover, this practice encourages the replenishment of subterranean water reservoirs and enhancements in the purity of water.

While these strategies do not fully restore the natural hydrologic cycle, they do significantly improve the urban environment. The Sustainable Urban Drainage System (SUDS) is currently regarded as the optimal collection of methods for managing water in urban areas.

SUDS are designed to allow water to either infiltrate into the ground or be retained in devices in order to mimic the natural disposal of surface water (Charlesworth et al. 2003; Charlesworth 2010). Among the goals of SUDS, are:

- Quantitative control of surface runoff;
- Improvement in the quality of water from surface runoff;
- Conservation of natural characteristics of bodies of water;
- Balance of hydrological variables in watersheds.

These four goals of SUDS contribute to flood control in urban watersheds, principally when they are still in the urbanization phase and the application of SUDS in new undertakings is possible.

This system is designed to manage the environmental risks resulting from urban runoff and to contribute wherever possible to environmental enhancement. SUDS objectives are, therefore, to minimize the impacts from the development on the quantity and quality of the runoff, and maximize amenity and biodiversity opportunities (Woods et al., 2007).

Uncertainty about long-term maintenance and other operational factors have retarded the widespread adoption of SUDS but environmental regulators and many local authorities and developers are keen to implement this approach in addition to traditional urban drainage systems (Andoh & Iwugo, 2002).

2. HISTORY OF SUDS

In order to comprehend the principles, suggestions, and imperative for the execution of Sustainable Systems, it is essential to provide a historical framework for the development of the urban drainage systems currently employed in our cities.

Through time, and up to the Modern Age, drainage works, as a rule, were not considered as necessary infrastructures providing conditions for the development and ordering of urban centers (Matos, 2003). Nevertheless, rainwater drainage systems have been found in much more ancient cities or city ruins.

In the period prior to the Christian Era, systems implemented by the Persians and Greeks are worthy of note. Drainage networks constructed by the Romans (8th Century B.C. ~ 3rd Century A.D.) may be observed yet today, with small sections still in operation. The same holds true for ruins of cities built by Pre-Columbian peoples in various countries of Latin America (TIM, 2008).

Before giving an account of the important transformation which occurred in the 19th Century related to sanitary waste systems and rainwater drainage, it is worthwhile to go back some years in history when, according to Silveira (2002), a period of elimination of flooded areas began, the burying of septic tanks and then their replacement by underground channeling systems. This process for removal of waste and rainwater began in Italy as a result of observing a correlation between mortality rates of people and animals and the sanitation system; after that, it was adopted in innumerable European cities as a public health measure.

Thus emerged the concept of sanitary and hygiene related drainage systems, with the principle of expelling waters from cities, whether rainwater or waste water, giving the false impression of a "problem solved". To understand why this solution did not completely solve the problem, one need only look at the resulting physical aspects and observe the immense erosion caused at the end of these drainage systems, as well as all the morphological changes imposed on receiving bodies of water.

Interestingly enough, rainwater drainage as a public measure did not evolve as a result of modernization of engineering practices in search of comfort, but rather as a recommendation of medical prophylaxis. Evidently, the task of making it concrete in public works and integrating it with the urban landscape belonged to engineers and urban planners, but, unfortunately, this only received greater impulse with the occurrence of cholera epidemics in large cities around the world in the 19th Century. Epidemics which occurred in Europe in the years 1832 and 1849 are of special note.

Between 1850 and the end of the 19th Century, many important cities of the world, principally European capitals, were provided with large single underground networks for waste materials (rainwater drainage and sewage conducted by the same conduits) (Silveira, 2002). According to Jones & MacDonald (2007), in terms of urban drainage, one of the most famous examples of this period was the reconstruction of Paris in the Second Empire of Haussmann and Napoleon III, who used a system designed to rapidly remove waters from the city.

The sanitary hygiene concept, as already mentioned, foresees rapid expulsion of waters from the city for the purpose of preserving the health of the population and eliminating any type of discomfort the water could cause. Nevertheless, what was not foreseen in this effort for channeling is the impact caused downstream, since the sanitary hygiene concept acts only in a local manner in the sense of transferring the problem to other regions. Thus, this concept, associated with rapid growth of the urban population, and by the latter, understanding all the characteristics brought about by urbanization such as impermeabilization of the soil and removal of plant cover, is responsible for constant flooding, landslides, and problems created in relation to recharging of aquifers.

Such a concept was adopted worldwide up to the 1960s of the past century (20th Century). In this period, developed countries began to perceive the conflict generated between the existing rainwater drainage system and the environment. Thus began a new drainage concept, an evolution of the former; concerned not only with public health, but also with the environmental question.

Over the past few decades, there has been a growing interest in studying sustainability, which has been referred to by different names in different countries. In the USA and Canada, it is known as Low Impact Development (LID), in the United Kingdom as Sustainable Urban Drainage Systems (SUDS), in Australia as Water Sensitive Urban Design (WSUD), and in New Zealand as Low Impact Urban Design and Development (LIUDD). Irrespective of the name, the sustainable systems discussed share similar principles and concepts, all of which emphasize the importance of maintaining a balance among the variables of the hydrologic cycle and their impacts on watersheds.

The benefits of this urban drainage treatment approach are strengthened by the positive impacts on landscaping, the environment, and the economy. It effectively manages not only the maximum flow rates, but also the overall volume, frequency, duration, and quality of runoff and drainage (Souza, 2005).

On the other hand, developing countries are lagging behind in terms of urban drainage management, as they have limited quantitative control over it and have not yet achieved satisfactory quality control of the drained water. This highlights the importance of investigating methods to promote the adoption of strategies, such as implementing fees for water usage, in order to achieve control at the origin and preservation of the attributes of the hydrologic cycle before human development.

3. HYDROLOGIC CONCEPTS USED IN SUDS

The Hydrologic Cycle encompasses the transformation of water as it transitions between its three distinct physical states: solid, liquid, and gas. The aforementioned factors exert a significant impact on the behavior of this cycle, particularly when considering a worldwide perspective.

On a local level, the amount of water and the rate at which it moves through the various stages of the hydrologic cycle are directly impacted by elements such as vegetation, elevation, land features, temperature, soil type and usage, and geological conditions.

The hydrologic cycle is a continuous process of water transport from the ocean to the atmosphere and back to the ocean through precipitation and runoff. Changes in the variables of this system can locally alter its characteristics.

In the realm of sustainable watersheds, it is crucial to uphold or counterbalance the primary factors (evaporation, evapotranspiration, precipitation, surface runoff, and subterranean flow) through direct interventions on one or more of these factors.

The hydrologic cycle operates through the process of evaporation, where water on the Earth's surface is heated by the Sun and rises into the atmosphere. Gravity induces the downward movement of condensed water, resulting in precipitation. After reaching the surface, water moves along streams (known as surface runoff) that collect into rivers and eventually flow into the oceans. Alternatively, it can seep into the earth through the pores, fissures, and cracks in soils and rocks, forming underground water flow.

However, not all water from precipitation will progress in this flow as a portion of it will be trapped in structures, trees, shrubs, and vegetation. Consequently, it does not reach the soil (evaporates) and is referred to as loss by interception.

Therefore, it may be stated that water that reaches the soil can follow different pathways; some of it will undergo evaporation and return to the atmosphere, while the rest will seep into the ground. However, when the intensity of rainfall exceeds the capacity for infiltration and evaporation, it results in the formation of small amounts of water known as depression storage.

When these depressions become completely filled and exceed their capacity, water initiates movement across the surface, a phenomenon referred to as surplus rainfall. When a layer of water forms and covers the path of flow, we refer to it as surface runoff. When the runoff is stored as it moves, it is referred to as detention storage.

However, it is important to note that the process is intricate and the other stages occur concurrently. As a result, some of the flow may seep into the soil or evaporate back into the atmosphere before reaching a water body. The water that seeps into the soil initially enters the zone of soil where plant roots are present. The upper layer of soil has the ability to hold a certain amount of water, which is referred to as field capacity.

Another pathway for the water that enters the soil is through direct evaporation into the atmosphere. This occurs when the water is taken up by plants and then released back into the atmosphere through transpiration. The process being referred to is known as evapotranspiration, which takes place at the uppermost section of the unsaturated zone. This zone is characterized by the presence of both air and water in the crevices between soil particles.

Therefore, processes may be distinguished as:

- **Evaporation:** this is the physical process in which a liquid or solid passes to the gaseous state due to solar radiation and to the processes of molecular and turbulent diffusion;
- **Transpiration:** this is the release of water vapor by plants by means of stomata and organelles located on leaves;
- **Evapotranspiration:** this is the sum of the evaporation and transpiration processes, given the difficulty of separating the two phenomena. Evapotranspiration is expressed in terms of water depth, in mm.

The main factors that interfere in the processes of Evaporation and Evapotranspiration are:

- **Air temperature:** the greater the temperature, the greater the capacity of the air for containing water vapor;
- **Water temperature:** the hotter the water, the greater the evaporation;
- **(Relative) atmospheric humidity:** the lower the relative air humidity, the greater the evaporative capacity of the air;
- **Wind:** the displacement of air masses constantly renews the mass of water vapor over the evaporating surface;
- **Salinity:** the presence of salts reduces evaporation (2% to 3%).

As water from earlier stages continues to seep into the earth, it enters the saturated zone and starts flowing underground, which helps replenish the aquifers by increasing the amount of stored water. The saturated zone, also known as the aquifer, is characterized by soil pores or fissures in rock formations that are completely filled with water, resulting in saturation. The uppermost region of the saturated zones aligns with the water table, making it particularly vulnerable to contamination.

Subterranean flow persists, and water may ascend to the surface via springs, nourishing bodies of water or discharging directly into the ocean.

The issue arises when human activities start to disrupt the aforementioned processes, and these modifications caused by humans on the ecosystem can potentially disrupt a portion of the hydrologic cycle.

From a global perspective, human activities that emit high levels of gases into the atmosphere contribute to the greenhouse effect, which in turn alters the conditions of thermal radiation emissions and disrupts the equilibrium of the system.

On a local scale, engineering activities typically lead to alterations in variables such as infiltration, evaporation, and runoff. These works include water engineering projects that alter the natural paths of rivers, lakes, and seas.

Furthermore, nearly all activities that engage in deforestation for the purpose of wood usage, or even for clearing the land for new construction projects, have a detrimental impact on the behavior of the watershed. Deforestation primarily leads to heightened runoff and decreased evapotranspiration, regardless of the kind of soil usage.

But of all human activities, urbanization is that which produces the greatest local changes in the processes of the Earth's hydrologic cycle. The most rapidly changed factor is the Runoff Coefficient, which is the ratio between the volume of surface water runoff and the volume of rainwater.

Various problems are caused by change in the Runoff Coefficient. These problems go beyond erosion of the drainage area; they cause imbalances in the rainwater channel (erosion and silting) and the ever so frequent urban flooding.

4. THE EFFECT OF URBANIZATION

Modifying the original course of a natural drainage basin, whether through the effects of forestry, agriculture, or urbanization, can result in significant alterations to the flow and retention of water (Booth, 1991).

Urbanization leads to the development of new engineering projects, including the construction of buildings, paving of streets, walkways, and the subsequent removal of natural vegetation. This alteration in the environment results in a change in the natural ability of these regions to allow water to pass through. As a result of this impermeabilization, there is a decrease in the ability of rainwater to seep into the ground, causing a significant rise in rainwater runoff.

Presently, in Brazil, every major urban center experiences instances of flooding, while smaller cities have at least a few areas that become swamped during periods of high water levels. This issue has progressively deteriorated over the years in Brazilian cities, with authorities struggling to effectively manage these occurrences.

This phenomenon primarily arises due to urbanization, which leads to the removal of pre-existing vegetation in watersheds and its replacement with impermeable surfaces such as walkways, paved streets, roofs, and parking lots.

These modifications ultimately result in alterations to the local hydrograph, resulting to a decrease in concentration times and an increase in peak flows. With an equivalent amount of rainfall, the runoff volume increases, leading to a higher likelihood of flooding in the lower or flat regions of the watershed.

Some engineering solutions bet on transfer of the problem by improving or increasing the rainwater drainage system, but in general, they end up transferring problems further downstream. In addition, upon transferring a greater volume of runoff to the final point of the drainage system, serious erosion problems are generated.

In addition, the presence of peak flows has resulted in increases of the frequency and seriousness of flooding and in intensification of erosion processes with increase of production, transport, and deposit of sediments. Impacts like these directly affect the quality of bodies of water in these watersheds.

Changes in the local hydrologic cycle and its consequences due to the urbanization process, as previously mentioned, directly affect the environmental balance and, thus, the quality of life of the resident population. Problems arise both due to the inflows to the system, which in future will be less certain, and increasing downstream hydraulic and regulatory constraints on outflows (Ashley et al., 2007).

In this context, it may be observed that the conventional rainwater drainage system with the principle of rapid flow downstream is not able to eliminate any disturbance that these waters may bring about.

5. TYPES OF SUSTAINABLE SYSTEMS

Among the sustainable systems that are being developed and implemented, the following devices or techniques may be highlighted:

- Permeable pavement;
- Semipermeable pavement;
- Detention and retention reservoirs;
- Infiltration trenches;
- Infiltration gullies;
- Infiltration wells;
- Microreservoirs;
- Rooftop reservoirs;
- Green roofing;
- Underground reservoirs; and,
- Grassed strips.

These devices, implemented in sustainable systems, aim to minimize the impacts of urbanization by reducing runoff and promoting substantial increases in infiltration rates. In order to achieve these outcomes, it is necessary to establish spaces that allow for the passage of substances and areas that retain precipitation. These areas should be integrated into a wider system that focuses on managing water resources.

These technologies can be utilized together or independently, depending on the specific project or local requirements and capabilities. This offers a favorable cost-to-benefit ratio, as well as social, economic, and environmental advantages. However, it is generally seen that systems yield a substantial outcome only when they are used together and throughout different sections of the watershed to form a larger project.

Currently, numerous countries are actively testing or implementing a wide range of sustainable drainage methods. In the following, we will outline the primary systems that can be utilized to manage urban watersheds in order to enhance or supplement traditional drainage systems.

5.1 Permeable pavement

The necessity of diverting surface waters away from their point of origin may be diminished or eradicated. This can be achieved by directing it to pass through permeable pavements constructed using materials such as concrete blocks, crushed rock, or porous asphalt. The water can either penetrate directly into the subsoil or be retained in an underground reservoir, such as a layer of crushed rock or a reservoir, depending on soil factors such as permeability and infiltration rates. If infiltration is not feasible or suitable (e.g., owing to soil pollution), an impermeable membrane can be employed as a barrier to prevent water from entering the pavement regardless of the circumstances. In these instances, the majority of pollutants are eliminated either on the surface or sub-base of the substance, or through the filtration process of the reservoir or subsoil.

Thus, it may be said that permeable pavement is an alternative infiltration device where surface runoff is diverted through a permeable surface into a rock reservoir located under the same surface (Urbonas & Stahre, 1993). According to the same authors, permeable pavement may be classified into three types:

- **Porous asphalt pavement:** has its upper layer constructed in a similar way to conventional pavements with a difference in the fraction of fine sand, which is removed from the mixture of the aggregates used in construction of the pavement;
- **Porous concrete pavement:** just as in porous asphalt pavement, it only has a difference in the sand fraction as compared to conventional pavements;

- Semipermeable pavement (described below).

For Silveira (2002), permeable pavements are pavements that normally act in control of the peak and volume of runoff, in control of diffuse pollution, and, when they infiltrate water in the soil, they promote recharging of underground waters. Porous pavements are appropriate for use on light traffic roadways, parking areas, pedestrian streets, squares, and sports courts.

Schlüter et al. (2002) studies have demonstrated that porous pavement can get good results, such as:

- Average percentage outflow of just under 50%;
- Lag times of 45 minutes for medium and 145 minutes for small events;
- Initial rainfall capacity in the order of 2.27 mm.

Abbott & Comino-Mateos (2003) have also shown the peaks of storm events were significantly reduced as they flowed out of the pavement system. For example, the peak outflow corresponding to a rainfall intensity of 12 mm/h was 0.37 mm/h. This is a major benefit of these systems because:

- It reduces the capacity of sewerage systems; and,
- Enables urban developments to comply more easily with discharge consents.

Even so, as mentioned above, this system only leads to temporary water storage and therefore needs to be joined with another system for retention and/or infiltration of the water absorbed, since this system does not have that function.

5.2 Semi-permeable pavement

For Urbonas & Stahre (1993), semi-permeable pavements are constructed of hollowed concrete blocks filled with granular material like sand or undergrowth, like grass, placed over a granular base layer and this layer is covered by geotextile filters to avoid migration of the granular base.

According to studies of Araújo *et al.* (2000), semipermeable pavements may be industrial concrete blocks, as well as paving blocks, among other items. According to Cruz *et al.* (1999), they have the function of providing for reduction in runoff volumes and in the time for recharging the watershed.

Moreover, according to Araújo *et al.* (2000), in their studies performed on different types of semipermeable pavements (concrete, concrete blocks, compacted soil, paving blocks, and hollowed blocks), after simulations of similar precipitation levels, the following runoffs were obtained, respectively: 17.45 mm, 15.00 mm, 12.32 mm, 10.99 mm, and 0.5 mm. These studies were able to prove the efficiency of different types of pavements in reduction of runoff through increase of infiltration rates.

Furthermore, according to Araújo *et al.* (2000), the use of permeable pavements eliminates the need for collection boxes and water conduits because the device practically does not generate runoff. In addition to implementation costs of permeable pavements, there are maintenance costs, which consist of cleaning the pores of porous pavements (porous concrete) with water jets and machines for vacuuming of dirt and sediment. These costs were not estimated due to the non-existence of companies specialized in maintenance of this type of device in the country. Nevertheless, to have an idea, average expense on maintenance in the United States is in the order of 1% to 2% of the implementation cost of the device.

5.3 Retention and detention reservoirs

Retention reservoirs offer a significant benefit in that they may be constructed in public spaces, such as squares, parks, and courts, which can still be used for various purposes even after rainfall. Detention reservoirs are equipped with a water layer and carefully regulated to ensure high water quality. They can be utilized in grassland marshes or urban reservoirs. Enclosed detention reservoirs have a cost that is seven times higher than that of open reservoirs. Another option is to use off-line systems, where a portion of the discharge flows laterally before returning to the main waterway (Parkinson et al., 2003).

Cruz et al. (1998) identified maintenance as the primary issue associated with detention reservoirs, as they impose significant responsibilities on their owner. This occurs because precipitation causes surface runoff to transport various solid residues and sediments present in the watershed. Hence, it is imperative to consider the requirement for acquiring devices that can perform regular cleaning of the reservoirs, thereby preventing any decline in their effectiveness.

The open device is particularly influenced by maintenance costs, as it requires more frequent upkeep due to public health concerns. Despite its lower implementation cost, this type of structure incurs an estimated annual maintenance cost of US$ 130.00. Consequently, it may take 4 or 5 years for the total cost to equal that of other systems (Cruz et al., 1998).

5.4 Wetlands

Wetlands are terrestrial regions where the soil retains a high level of water, either consistently or during specific periods of time. These regions can be either natural or man-made, and they can be partially or completely submerged under a specific water depth. This wetlands system include several types of wetlands, such as swamps, marshes, and lowlands. Furthermore, wetlands can contain water that is either saline, freshwater, or brackish.

While wet or dry lagoons can be built, incorporating permanent water enhances their visual appeal and supports biodiversity. Furthermore, while managing a flooded system, it is possible for them to be engineered to handle significant fluctuations in water levels caused by precipitation, so enhancing their ability to store water and protect against floods.

Another crucial attribute is that the extent of solids removal can be substantial when there is sufficient retention time. The algae and plants found in wetlands exhibit a high level of efficacy in filtering and eliminating nutrients. Lagoons and wetlands can receive water from drainage gullies or piping systems, as well as minor amounts of municipal sewage. Additionally, they function as sediment "traps" to help manage sedimentation issues.

Costa et al. (2003) suggest that created wetlands are an effective and affordable way for decreasing the dangers associated with residual waters and lowering microbial contamination. These wetlands utilize simple technology that is easy to run. Within them, there is primarily efficient cycling of nutrients, elimination of organic matter, and reduction of pathogenic bacteria found in residual waters. The main mechanisms responsible for this removal include settling (caused by microbial biofilm adhering to the roots and substrate), predation, competition among microorganisms, and the potential release of toxic substances by plants through their roots (Brix, 1994). According to Zheng et al. (2006), using a combination of wetlands and infiltration ponds is a cost-effective method for managing drainage at the end of a pipe. This approach can be used for controlling local sources of water as part of urban growth and regeneration.

In summary, it may be said that these systems have important functions, among which the following stand out:

- Their capacity for regularization of water flows, moderating peaks of flooding (according to their designed capacity);
- Capacity for modifying and controlling water quality (in the event of limits to be established);
- Their importance in the function of reproduction and feeding of aquatic fauna;
- A form of protection of local biodiversity as a refuge area for land fauna;
- A silting control system for rivers due to their capacity in retaining sediments.

Even with these benefits, this system may be adopted only after adequacy studies in regard to characteristics of the location, as well as studies for avoiding problems with proliferation of insects that may cause health risks to those living near the area.

5.5 Infiltration trench

Infiltration trenches are reservoirs filled with rocks that are designed to collect rainwater for temporary storage, allowing the water to slowly seep into the soil over time. The lifespan of the system can be extended by adding a filter or deposit that eliminates the excessive entry of solids, preventing the system from becoming clogged.

Silva (2007) defines infiltration trenches as linear constructions with a length that is predominantly greater than their breadth and depth. These trenches are designed for percolation and/or draining purposes. Their geometry is contingent upon the permeability of the soil and the surface area that is accessible for infiltration to take place. The design may prioritize infiltration, storage, or both, depending on local conditions and the volume to be infiltrated. Trenches are typically designed to accommodate significant amounts of water infiltration, are enclosed, and can be utilized for landscaping purposes in conjunction with other constructions.

According to Silveira (2008), the main function of infiltration trenches is to reduce runoff and promote recharging of aquifers, but another important function is to promote water treatment by means of soil infiltration.

Waste filters (also known as French drains) are widely used by authorities along highways. They are similar to infiltration trenches, but use a perforated pipe, which carries the flow along the gulley. This allows storage, filtration, and infiltration of water that passes from the collection area to the final point. Pollutants are removed by absorption, filtering and microbial decomposition in the surrounding soil. That way, these systems may be successfully designed so as to incorporate infiltration as well as to serve as efficient filtration systems.

Thus, according to Nascimento (1996), the advantages in the use of this type of structure are:

- Reduction or even elimination of the local microdrainage network;
- It avoids reconstruction of the downstream network in the case of saturation;
- Reduction of risks of flooding;
- Reduction in pollution of surface waters;
- Recharging of underground waters;
- Good integration with the urban area.

The cost of implementation of an infiltration trench, according to Souza & Goldenfum (1999), basically depends on the cost of earthwork and the cost of materials (crushed rock and geotextile). A trench that drains an area of around 300

m^2 costs around US$ 239.00 (experimental module). That gives us an estimated cost of US$ 0.79/m^2.

Infiltration systems may also be very effective in industrial zones. But the use of solutions based on infiltration for these locations requires careful reflection, principally because there is the risk of environmental damages caused by soil contamination. In this case, the focus will be on not mobilizing the contaminants.

5.6 Infiltration ditch and gulley

Ditches and gullies are compensatory techniques consisting of simple depressions dug into the ground for the purpose of collecting rainwater, making temporary storage and favoring infiltration (Silva, 2007).

The depressions are linear and on permeable land, where generally there is a grassy covering. In addition, the infiltration gulley may incorporate small deceleration dams that favor infiltration and protect against erosion. Nevertheless, according to Silveira (2002), this system may cause colmatation and allow the passage of pollutants and it has a propensity to retain stagnant water. Therefore it needs frequent maintenance.

Ditches are characteristically works of large width and low slope in the longitudinal direction dug into the earth. Gullies, for their part, are ditches that are not very deep (Brito, 2006). In both cases, they may also be used as one of the management resources within gardened areas of the urbanized watershed, or may even be incorporated in leisure areas.

These systems provide temporary storage for rainwater, reducing peak flows of waters and also aid in filtering of pollutants (deposited in the impermeable areas). They are often installed as part of a drainage network, where they help to connect it to a lagoon or wetland before final discharge into a natural waterway. Even so, they have often been installed along avenues, streets, and roadways to substitute conventional drainage systems.

5.7 Microreservoir

For Loganathan *et al.* (1985), microreservoirs are storage devices for precipitated water which act in the sense of retarding concentration time, mitigating the peaks of flow hydrographs.

Agra (2001) defines it as simple structures in the form of boxes similar to those used for supply water. They may be made of diverse types of materials as, for example, concrete, masonry, PVC, or another material, having a discharge structure like an orifice. They are normally buried, but they may be open to view provided that there is a height limitation due to the drainage network.

According to Cruz *et al.* (1998), the general cost of implementation of an underground microreservoir for control in urban lots is, on average, from US$ 400.00 to US$ 500.00, with the need for additional spending on its maintenance.

5.8 Infiltration well

For Reis *et al.* (2008), infiltration wells are isolated devices that allow infiltration of runoff into the soil. They may be structured through filling with crushed rock (porous medium), as well as being lined by perforated concrete piping or bricks laid in a staggered manner surrounded by a geotextile sheet making the soil/piping interface. They have the advantages of low cost of execution and attempt at balance of the urban hydrologic cycle by intermediation of recharging the water table.

Nevertheless, for Silveira (2002), the well also has its drawbacks for not having the capacity for supporting large loads of sediments and offering risks in regard to infiltration of pollutants.

According to Reis (2005), the cost of execution of an infiltration well for an area of 500 m^2, determined in December 2004, was US$ 1,312.50.

5.9 Rooftop reservoir

According to Silva (2004), this is a device that seeks to compensate the effect of impermeabilization by means of the impact structure itself, which in this case is the roof. According to Azzout *et al.* (1994), the rooftop reservoir functions as provisional storage of rainwater, releasing it gradually to the rainwater network by means of a specific regulating device. For Silva (2004), both roofing tiles as well as concrete roofing structures that can store rainwater may also be used for the purpose of collecting water.

Nevertheless, Azzout *et al.* (1994) present considerations in regard to the rooftop reservoir, expressing the need to take care with more highly sloped rooftops by the fact that its installation is restricted to rooftops with a slope limit of 2%, also emphasizing the difficulty of installing the reservoir on already existing rooftops and possibly the high cost of the device.

With respect to its pollutants, roof runoff is generally considered as non-polluted, or at least not significantly polluted, compared to waste waters and highway runoff, since it consists of rainwater flowing over various, in general less abrasive materials, such as tiles, bitumen, less corrosive metals and concrete. In general, it may be stated that the quality of runoff may be ameliorated or become worse compared to the composition of atmospheric deposition depending on the kind of materials used for roof cover and drainage systems and their interaction with the atmospheric deposition. For building materials displaying open air corrosion data or data of leaching experiments, runoff loads can even be assessed (Zobrist et al., 2000).

5.10 Green roofing

Other techniques that reduce the flow of rainwater and improve water quality include green roofing and its reuse. Green roofing may reduce local peak flows and the total volume discharged into the conventional drainage system. In addition, they may improve thermal and acoustic insulation and increase the useful life of the roofing.

Reuse of rainwater involves collecting it at the location (residential, commercial or industrial) and its use as a substitute for treated water (with high treatment and distribution costs), for example in watering the garden, washing sidewalks, or even for flushing of toilets. Reuse of rainwater will be dealt with in the following item.

According to Heneine (2008), green roofing is currently most widespread in German speaking countries of Central Europe and is spreading to the north and northwest of Europe and North America. Such devices consist of a green covering composed of plants and soil, emphasizing that the plant cover has average growth and is planted on an impermeable base. However, additional layers, as for example, a barrier of roots, and a drainage and irrigation system may and should be included.

Heneine (2008) affirms that green roofing provides for reduction of the effects of torrents of rain, carry benefits for fauna, and mitigate heat in buildings. In relation to planting style, there are two distinct focuses, one being "intensive" (which needs more soil, more depth, and accommodates larger plants that may be as large as trees and bushes) and the other "extensive" (which needs little soil and contains vines and grasses).

This technology arose as a new proposal for retaining precipitated water through the natural holding power existing in the vegetation and also generating delay in runoff from the system. That way, compared with unprotected roofing that has no type of water containment, the volume generated from a roof with the green roofing technology is less and with a greater concentration time in relation to urban drainage systems.

Studies undertaken by Oliveira (2009) show that in addition to its precipitation retaining characteristics and its delaying of runoff, green roofing presents thermal comfort characteristics. It is very efficient in reducing the temperature of the internal and external environment of buildings, even modifying the surrounding microclimate and mesoclimate because this roofing reduces heat emission.

Confirming the thesis that green roofing relieves heat in buildings, this system can lead to a reduction of around 25% in cooling needs, and with a substrate layer of 12 cm, it can reduce outside noise by around 40 decibels, and a substrate layer of 20 cm can reduce it from 46 to 50 decibels.

In general, green roofing makes use of three systems:

- The Alveolar System which permits the use of a greater variety of plants, including native species (ecological system); in this case a roof of grasses may be used because the alveoli retain a greater quantity of water;

- The Planar System characterized by the use of a layer of water over a raised base made of support modules, increases the benefits of retaining precipitated water, and also the thermal comfort of the area below it;
- The Modular System which is composed of prefabricated modules (already planted) where these modules are placed one beside another over an anti-rooting membrane and a membrane for retaining of nutrients (support systems of the modules).

5.11 Reuse of rainwater

For Fendrich (2003), it is worth emphasizing that all captured rainwater will reduce the waters that may cause flooding. Thus, this application is highly important for controlling water balance and it may play an important social role in areas subject to drought.

When there is the occurrence of rainfall and there is no system for capturing this water, it will end up in storm drains. Until this water arrives at the storm drain, it traces a route where it comes in contact with innumerable type of residues, contaminating it and making it inappropriate for consumption. Nevertheless, this water may potentially be used in many day-to-day situations. According to Pereira (2008), through Brazilian legislation, all water derived from rains is treated as sewage, creating the need for this water to be treated before its use, even without having been used.

5.12 Underground reservoir

According to Silveira (2002), an underground or buried reservoir is a type of impervious tank built below the ground (with impermeable concrete walls) allowing utilization of the surface for another purpose. In general, an underground reservoir operates as an impermeable open air detaining reservoir. Therefore, it weakens the runoff introduced in it through the controlled effect of water release through the orifice and valve at the bottom. Nevertheless it should have a mechanism to avoid the accumulation of pollutants and sediment, it requires frequent maintenance, and it is restricted to areas with more frequent rainfall.

5.13 Grassed strips

According to Silveira (2002), grassed or tree-lined strips are conceived to decelerate and partially infiltrate laminar flow coming from impermeable urban surfaces (parking areas and other surfaces), but may have its application associated with other situations. In macrodrainage, they assume the role of escape zones for flooding. Therefore, the main benefit of grassed strips is that in addition to significantly reducing the speed of surface runoff, they significantly help in reduction of peak flows in urban areas (when applied in large stretches).

Another focus could be given regarding the use of grassed strips. The concentration of sediment decreased exponentially with distance along the grass strip, reaching sometimes a constant value. Particles less than 5.8 μm had a very low trapping efficiency (sometimes negligible), while a substantial proportion of particles above 57 μm were trapped. Knowing that fine particles are the main source of pollution, this implies that grass is very effective for sediment control, but is less efficient in removal of other pollutants. The particle trapping efficiency clearly depended on grass length and particle size. To a some extent, it also depended on sediment concentrations in inflow (Deletic, 2005).

However, some reservations must be made, as for example the need for more constant maintenance.

6. CONCISE SUMMARY

Similar to other drainage systems, the SUDS system possesses both benefits and drawbacks. Consequently, both favorable and unfavorable aspects might be mentioned in relation to its utilization, which can be valuable for guiding future investigations and implementations.

Among the various devices discussed in this book, some notable advantages become apparent, including:

- Increase in the infiltration rates and reduction of surface runoff;
- Retaining of rainwater for later use in less refined activities (watering gardens, flushing toilets in bathrooms, washing sidewalks, etc.);
- Creation of leisure areas and a better landscaping aspect for cities.
- Nevertheless, the challenges that must be overcome for improvement in performance and incentive for application of these systems must be kept in mind. Among the main problems are:
- Need for frequent maintenance;
- High cost of implementation when necessary adaptations in pre-existing systems are made;
- In general, these systems do not support high loads of sediments and present the risk of colmatation.

However, when considering the implementation of some component systems of SUDS, the cost is not excessive. Therefore, after conducting an analysis of the economic and environmental feasibility, SUDS demonstrates a favorable cost-benefit ratio.

Society and its systems evolve through two primary mechanisms: initially, the pursuit of comfort and improved quality of life, and subsequently, the essential drive to ensure the preservation of life. Therefore, much as the management of rainwater through channels was crucial for preventing diseases and epidemics, a new drainage system is required to address conflicts arising from the interaction between urbanization and the Earth's water cycle.

Despite its relative novelty compared to traditional urban drainage systems, Sustainable Urban Drainage Systems (SUDS) have already been deployed in several locations as an alternate method for improving urban drainage. SUDS efficiently reduce peak flows caused by human activities. While SUDS is not widely implemented in Brazil, it is being extensively researched and utilized in several industrialized countries, particularly in Scandinavia and the United Kingdom.

Thus, it can be inferred that the implementation of Sustainable Urban Drainage Systems (SUDS) represents a progressive step in drainage systems, leading to an enhancement in the overall well-being of the affected population. It revitalizes the quest for creating sustainable settings in urban areas.

REFERENCES

Abbot, C. L. & Comino-Mateos, L. 2003. In-situ hydraulic performance of a permeable pavement sustainable urban drainage system. Journal of the Chartered Institution of Water and Environmental Management, v.17, n.3, p. 187–90.

Agra, S. G. Estudo Experimental de Microrreservatórios para Controle do Escoamento Superficial. Dissertação submetida ao Programa de Pós-Graduação em Recursos Hídricos e Saneamento Ambiental da Universidade Federal do Rio Grande do Sul, 2001. 104p

Andoh, R. Y. G. & and Iwugo, K. O. 2002. Sustainable Urban Drainage Systems: - A UK Perspective. 9th International Conference on Urban Drainage (9 ICUD), Portland, Oregon

Araújo, P. R.; Goldenfum, J. A. & Tucci, C. E. M. Avaliação da eficiência dos pavimentos permeáveis na redução de escoamento superficial. In: Revista Brasileira de Recursos Hídricos. v.5, n.3, 2000. p. 21-29

Ashley, Richard; Blanksby, John; Cashman, Adrian; Jack, Lynne; Wright, Grant; Packman, John; Fewtrell, Lorna; Poole, Tony & Maksimovic, Cedo. 2007. Adaptable Urban Drainage: Addressing Change in Intensity, Occurrence and Uncertainty of Stormwater (AUDACIOUS). Built Environment, v.33, n.1, p. 70-84

Azzout, Y.; Barraud, S.; Cres, F. N. & Alfakih, E. Techniques alternatives en assainissement pluvial. Paris: GRAIE, 1994

Booth, D. B. Urbanization and the natural drainage system – impacts, solutions, and prognoses. The Institute for Environmental Studies, University of Washington: The Northwest Environmental Journal, v.7, n.1, 1991.

Brito, D. S. Metodologia para seleção de alternativas de sistemas de drenagem. Dissertação de Mestrado em Tecnologia Ambiental e Recursos Hídricos, Publicação PTARH.DM- 094/06, Departamento de Engenharia Civil e Ambiental, Universidade Brasília, Brasília, DF, 2006. 117p.

Brix, H. Functions of macrophytes in constructed wetlands. Water Sci. Tech., v.29, n.4, 1994. p.71-78

Charlesworth S. 2010. A review of the adaptation and mitigation of Global Climate Change using Sustainable Drainage in cities. Journal of Water and Climate Change, v.1, n.3, pp. 165-180

Charlesworth S. M.; Harker, E. & Rickard, S. 2003. Sustainable Drainage Systems (SuDS): A soft option for hard drainage questions?. Geography, v.88, n.2, p. 99-107

Costa, L. L.; Ceballos, B. S. O.; Meira, C. M. B. S. & Cavalcanti, M. L. F. 2003. Eficiência de Wetlands construídos com dez dias de detenção hidráulica na remoção de colífagos e bacteriófagos. Revista De Biologia E Ciências Da Terra, v.3, n.1

Cruz, M. A. S.; Araújo, P. R. & Souza, V. C. B. 1999. Estruturas de controle do escoamento urbano na microdrenagem, XIII Simpósio Brasileiro de recursos hídricos, Belo Horizonte, 21p

Cruz, M. A. S.; Tucci, C. E. M. & Silveira, A. L. L. Controle do escoamento com detenção em lotes urbanos. RBRH – Revista Brasileira de Recursos Hídricos. v. 3, n. 4, 1998. p. 19-31

Deletic, A. 2005. Sediment transport in urban runoff over grassed areas. Journal of Hydrology, v. 301, issues 1-4, p. 108–122

Fendrich, R. Aplicabilidade do armazenamento utilização e infiltração das águas pluviais na drenagem urbana. 17 f. Tese (Doutorado em Geologia) – Programa de Pós-Graduação em Geologia, Universidade Federal do Paraná, Curitiba, 2003

Heneine, M. C. A. S. Cobertura Verde. Monografia apresentada ao Curso de Especialização em Construção Civil. Escola de Engenharia da UFMG. Belo Horizonte, 2008. 49 p

Jones, P. & Macdonald, N. Making space for unruly water: Sustainable drainage systems and the disciplining of surface runoff. Geoforum, n. 38, 2007. p. 534-544

Loganathan, V. G.; Delleur, J. W. & Segarra, R. I. Planning detention storage for stormwater management. Journal of Water Resourses Planning and Management. ASCE. New York, v. 111, n. 4, 1985. p. 382-398

Matos, J. S. Aspectos Históricos a Actuais da Evolução da Drenagem de Águas Residuais em Meio Urbano. Revista Engenharia Civil, Lisboa, n. 16, 2003. p. 13-23

Nascimento, N. O. 1996. Curso: Tecnologias Alternativas em Drenagem Urbana. Escola de Engenharia - UFMG

Oliveira, E. W. N. Telhados verdes para habitações de interesse social: retenção das águas pluviais e conforto térmico. Dissertação de mestrado. Rio de Janeiro: UERJ, 2009. 87p

Parkinson, J.; Milograna, J.; Campos, L. C. & Campos, R. Drenagem Urbana Sustentável no Brasil. Relatório do Workshop em Goiânia-GO, 2003

Pereira, L. R. Viabilidade Economico/Ambiental da implantação de um sistema de captação e aproveitamento de água pluvial em edificação de 100m2 de cobertura. Monografia (Trabalho de Conclusão de Curso) - Pontifícia Universidade Católica de Goiás, 2008

Reis, R. P. A. Proposição de parâmetros de dimensionamento e avaliação de desempenho de poço de infiltração de água pluvial. Dissertação (Mestrado em Engenharia Civil) - Universidade Federal de Goiás, Goiânia, 2005. 228p

Reis, R. P. A.; Oliveira, L. H. & Sales, M. M. Sistemas de drenagem na fonte por poços de infiltração de águas pluviais. Associação Nacional de Tecnologia do Ambiente Construído. v. 8, n. 2, 2008. p. 99-117

Schlüter, W.; Spitzer, A. & Jefferies, C. Performance of Three Sustainable Urban Drainage Systems in East Scotland. 9th International Conference on Urban Drainage (9 ICUD), Portland, Oregon

Silva, J. P. Estudos Preliminares para Implantação de Trincheiras de Infiltração. Dissertação de Mestrado em Geotecnia. Universidade de Brasília, 2007

Silva, L. C. Sistemas de Drenagem Urbana Não-convencionais. Trabalho de Conclusão de Curso. Universidade Anhembi Morumbi. São Paulo, 2004

Silveira, A. L. L. Apostila: Drenagem Urbana: aspectos de gestão. 1. Ed. Curso preparado por: Instituto de Pesquisas Hidráulicas, Universidade Federal do Rio Grande do Sul e Fundo Setorial de Recursos Hídricos (CNPq), 2002

Silveira, G. L. Cobrança pela Drenagem Urbana de Águas Pluviais: incentivo à sustentabilidade. Relatório de Pós-Doutorado, 2008

Souza, C. F. Mecanismos técnico-institucionais para a sustentabilidade da drenagem urbana, 2005, 174f. Dissertação (Mestrado) – Universidade Federal do Rio Grande do Sul. Programa de Pós-Graduação em Recursos Hídricos e Saneamento Ambiental, Porto Alegre, 2005

Souza, V. C. & Goldenfum, J. A. 1999. Trincheiras e infiltração como elemento de controle do escoamento superficial: um estudo experimental. XIII Simpósio Brasileiro de Recursos Hídricos. Belo Horizonte, 11p

TIM-II - Trabalho de Integralização Multidisciplinar II. Universidade Federal de Minas Gerais, Escola de Engenharia, Curso de Engenharia Civil, Projeto de Infra-Estrutura e Equipamentos Urbanos: Termo de Referência, 2008. p.25

Urbonas, B. & Stahre, P. 1993. Stormwater Best management Practices and Detention. Prentice Hall, Englewood Cliffs, New Jersey. 450p

Woods, B. B.; Kellagher, R.; Martin, P.; Jefferies, C.; Bray R. & Shaffer, P. 2007. The SUDS manual (C697). Construction Industry Research and Information Association, London. 697p

Zheng, J.; Nanbakhsh, H. & Scholz, M. 2006. Case study: design and operation of sustainable urban infiltration ponds treating storm runoff. Journal of Urban Planning and Development – Am. Soc. Civ. Eng. 132, pp. 36–41

Zobrist, J.; Müller, S. R.; Ammannm, A.; Bucheli, T. D.; Mottier, V.; Ochs, M.; Schoenenberger, R.; Eugster, J. & Boller, M. 2000. Quality of roof runoff for groundwater infiltration. Wat. Res., v.34, n.5, p. 1455-1462

04 TYPICAL ESTIMATIONS OF WATER INFLOW INTO TUNNELS

1. INTRODUCTION: AN OVERVIEW

Accurately estimating the input of groundwater into tunnels in hard rock is a challenging task. Practically, predictions vary from significantly underestimated, resulting in substantial cost overruns and unsafe working conditions, to significantly overestimated, causing contractors to disregard them. This chapter examines the factors that contribute to the challenges associated with estimating inflow. Additionally, it proposes several pragmatic approaches to overcome some of these challenges.

There are two primary issues associated with inflow estimations. One issue is the absence of uncomplicated, practical equations or models that may be easily used for hard-rock tunnels. This challenge may prove to be inevitable, and this chapter does not make an effort to enhance the situation. The second challenge lies in the fact that the actual permeability range in fractured rock usually spans at least six orders of magnitude. Furthermore, this range tends to repeat multiple times throughout the lengths of long tunnels. The unique combination of permeability range and tunnel length distinguishes hard-rock tunnels from well fields, large aquifers, and other practical hydrology applications.

This chapter focuses on the excavation of hard-rock tunnels in fractured rock that are located below the groundwater table. These tunnels are typically built under normal atmospheric pressure using either drill and blast techniques or main-beam tunnel boring machines. The installation of lining in these tunnels is normally done post-excavation, and only in areas where it is necessary. Throughout the construction process, groundwater infiltrates the tunnels unhindered via fractures in the rock. In areas where the rock is impermeable and the pressure of the water above the tunnel is low, the amount of water flowing into the tunnel will be little. In areas where the rock has significant, unobstructed cracks or where the water pressure is great, there will be a significant amount of water flowing in. In locations where the rock exhibits both extensive fissures and significant hydraulic pressure, the influx of water can have devastating consequences.

Tunnel designers are responsible for assessing and communicating the anticipated water volume that will be encountered over the whole length of the tunnel, as well as in the specific location where the excavation is taking place. The total flow is utilized to adequately size pumping systems and water treatment plants. The incoming flows in the heading will have an impact on the contractor's building methods and timetable. Significant delays can arise if either one is underestimated. If either the cost is excessively overstated, it might lead to excessive and unneeded expenses.

Within hard-rock tunnels, the majority of the inflow originates from a limited number of locations, but a portion of the inflow is sourced from numerous locations. Additionally, a significant portion of the tunnel remains devoid of moisture. The cumulative inflow increases along the length of the tunnel and represents the total of all the inflows. This finding is the basic foundation of hard rock tunnels and is the main cause of difficulties in accurately measuring inflow using conventional hydrologic methods.

Various analytical expressions have been proposed in literature to compute groundwater discharges into tunnels, including the works of Goodman (1965), Lohman (1972), Zhang (1993), Heuer (1995), Lei (1999), Karlsrud (2001), Raymer (2001), and El Tani (2003). Furthermore, Katibeh and Aalianvari (2010) introduced a novel approach for categorizing tunnel length based on the infiltration of groundwater into the tunnels. This chapter presents an overview of analytical techniques for estimating the amount of groundwater flowing into tunnels.

2. INFLOW EQUATIONS

The equations for groundwater influx are derived from Darcy's Law and the principle of mass conservation. Darcy's Law states that the rate of fluid flow is equal to the product of permeability and the gradient. The principle of mass conservation states that the amount of water entering a system is equal to the sum of the water replenished and the water discharged from storage.

2.1 Thiem equation

Darcy's Law is Q = (KA)I, where I is the hydraulic gradient, KA is the permeability (K) across an area (A), and Q is the inflow. In this version of the equation (KA) are taken together in parentheses as a single term to emphasize the fact that actual water flows through a finite volume of rock mass, rather than a theoretical unity.

If Darcy's Law is configured for cylindrical coordinates around a vertical well, then the Thiem equation from well hydraulics results (Figure 1):

$$Q = 2\pi T\,(H_2 - H_1) / \ln(r_2/r_1) \qquad (1)$$

where the gradient is expressed as $(H_2-H_1)/\ln(r_2/r_1)$. H_1 and H_2 are the potentiometric heads in the aquifer at two arbitrary points having radial distances r1 and r2 from the center of the well. The hydraulic conductivity (K) and unit area (A) are handled together in the term 2pT, where T is the transmissivity. Provided that the rock is uniformly permeable, then T=Kb, where b is the vertical thickness of the aquifer. The Thiem equation (as shown above) is based on the following assumptions:

- Flow is radial toward the well and non-turbulent. The well produces from the full thickness of the aquifer.
- The well has reached steady-state flow, meaning that the cone of depression around the well has encountered a supply of water sufficient to replenish the water produced from the well. At this point, the cone of depression stops expanding and water is no longer being released from storage.

The Thiem equation can be applied to tunnels by turning the well on its side. The axis of the well is now the centerline of the tunnel. As with a well, the cone of depression is radial around the tunnel. Unlike a well, the potentiometric surface is a bit more abstract, and will look like a trough along the axis of the tunnel if measured at springline (or any constant elevation).

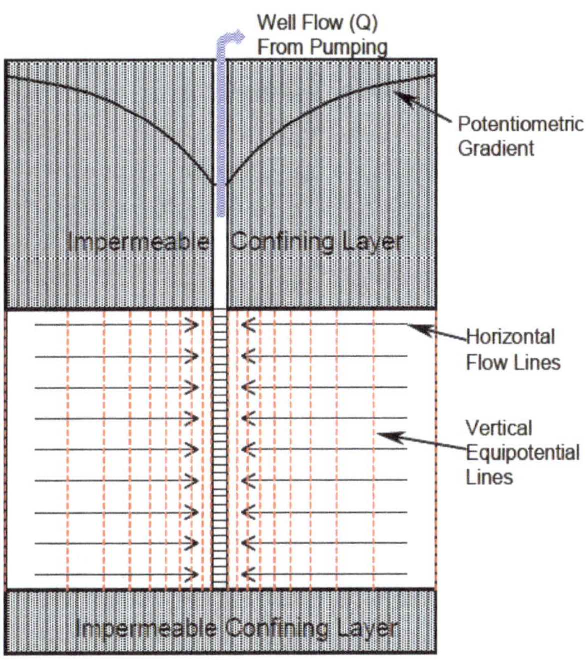

Fig. 1. Thiem equation for radial flow to wells. Equipotential lines are vertical and concentric about the well. Flow lines are horizontal and radial toward the well. The aquifer extends infinitely, and at infinity recharge equals well flow.

For convenience, r_1 can be the tunnel radius and H1 the potentiometric head at the tunnel radius. If the tunnel is under construction at atmospheric pressure, H_1 is the elevation of the tunnel wall. H_2 is the head of water some distance r_2 from the centerline of the tunnel. Transmissivity, which is the coefficient of proportionality, now has to be oriented horizontally along the axis of the tunnel, becoming the "horizontal transmissivity."

Note that H1 is not a specific point, but ranges between the crown and invert of the tunnel. This difference is trivial for deep tunnels. For large-diameter, shallow tunnels, however, this difference might become significant.

"Horizontal transmissivity" (T_h) is the coefficient of proportionality for the horizontal Thiem equation. It is placed in quotes because transmissivity is normally considered to represent the vertical thickness of the aquifer, rather than the horizontal length of the tunnel. The concept is important and worthy of indulgence to make a point. The fundamental observation shows that permeability varies over many orders of magnitude along the length of a hard-rock tunnel. This is similar to the horizontal stratification in a sedimentary aquifer penetrated by a vertical well. The "horizontal transmissivity" requires that the average permeability (Kavg) be considered when calculating inflow, rather than the idealized value of hydraulic conductivity. The average hydraulic conductivity is Kavg =Th/L, where L is the length of the tunnel.

The assumptions of the Thiem equation still have to apply for tunnels. Flow has to be radial toward the tunnel. The tunnel has to be long, such that that non-radial flow around the ends of the tunnel is negligible compared to the total inflow. The water table has to be high above the tunnel and not draw down close to the tunnel. The cone of depression has to have room to expand radially to a point where the water captured is large enough to provide for the inflow, even if an infinite supply of water is never encountered. These assumptions are reasonable for tunnels as long as the rock is nearly tight, the inflows very small, and the tunnel deep below the water table. In practice, tunnels that meet these conditions are less likely to have groundwater problems becausc the inflow rates will have to be very small.

2.2 Goodman's equation

Goodman (1965) considered the question of a tunnel lying beneath a lake or large river. He considered this lake or river to be an infinite source of water and applied the method of images (Lohman, 1972) to derive the following equation (Figure 2):

$$Q_L = 2\pi K H_0 / \ln(2z/r) \qquad (2)$$

where H0 is the head of water above the tunnel, and z is the distance from the tunnel to the bottom of the lake. Goodman also divided T by the length of the tunnel to put the equation in terms of hydraulic conductivity (K) and inflow per unit length of tunnel (QL). This equation only applies to steady-state inflow along the length of the tunnel. The inflow is steady state because the lake acts as an infinite recharge boundary, which causes the cone of depression to stop expanding.

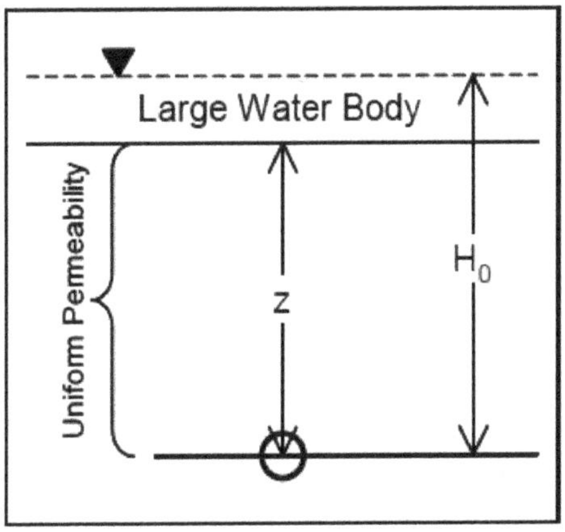

Fig. 2. Goodman's (1965) model for tunnels beneath a large water body. z is distance from centerline to top of rock. H0 is initial head between water surface and centerline.

Workers who have tried to apply this equation commonly report that actual inflows deviate greatly from the inflows predicted by this equation. Zhang and Franklin (1993) report that measured inflows range from 90 percent lower to 30 percent higher than predicted by Eq. 2. Heuer's (1995) work indicates that inflows from his projects have tended to be about one eighth (87.5 percent) lower than predicted by Eq. 2. My own experience from fairly shallow tunnels (typically less than 100 m) is in rough agreement with Heuer's. Other colleagues of mine report that inflows from deeper tunnels increase toward and then pass the inflows predicted by Eq. 2 as the tunnels become deeper.

Freeze and Cheery (1979) further modified Goodman's (1965) solution by replacing z with H0. In this solution, the water table is modeled as an infinite recharge boundary. Freeze and Cherry's (1979) solution will give minimally lower estimates than Goodman's because H0 is greater than z and both terms are in the logarithm. In a real tunnel, however, it is unlikely that recharge to the water table from precipitation could keep up with the yield from the larger fractures. This will serve to draw down the water table substantially around the tunnel and thus lower the head (H0) to a point where recharge can keep up with the reduced inflow.

2.2.1 Assumptions

Goodman's model relies on several significant simplifying assumptions. Initially, it should be noted that the tunnel is of an indeterminate length and the water table will never be much lowered at the highest point of the tunnel. Furthermore, at a certain elevation z above the tunnel, there exists an inexhaustible water source, such as a lake or river, which cannot be emptied by the water flowing into the tunnel. The elevation of the head (H0) must be equal to or greater than the elevation of the base of this reservoir (z) above the tunnel. Furthermore, the hydraulic conductivity (K) of the earth separating the tunnel and the base of the reservoir is the sole determinant of the speed at which the unlimited water supply flows into the tunnel. Furthermore, the fluid motion is characterized by a lack of turbulence, and the ground is uniform and exhibits the same properties in all directions, such as a consistent composition of sand or silt without any fractures.

Goodman's assumption of an inexhaustible source of rock material above the tunnel is likely to be somewhat applicable in several hard-rock tunnel scenarios in the eastern United States, such as the Chattahoochee Tunnel. The reservoir does not need to be specifically a lake or a river; it simply has to be a dense, saturated area that is significantly more porous and permeable than the bedrock that holds the tunnel. The saturated zone above must receive an excess amount of recharge from rivers, rainfall, or overlaying formations in order to compensate for any leakage that occurs through the bedrock zone and into the tunnel. For the purpose of this analysis, the transition and soil zones of the Chattahoochee Tunnel are deemed sufficient to fulfill this requirement.

Goodman's assumption that the aquifer is homogenous and isotropic is not feasible for hard-rock tunnels. However, Heuer's empirical reduction factor of 1/8 is assumed to compensate for this. Although the exact calculation of Heuer's reduction factor has not been determined, it appears to be in line with the decrease in flow anticipated by fracture-flow calculations, rather than porous-media equations like Goodman's. Additional research and efforts are required in this field.

2.2.2 Practical ranges of variables

The variables in Eq. 2 have practical ranges for tunnels. These practical ranges give insight into which are more important and which are less. In summary, K is the most important term and hardest to estimate, H0 is less important and easy to estimate, and ln(2z/r) is of minor importance and easy to estimate.

H0: In most situations, the maximum for the static head (H0) is the elevation difference between the tunnel and highest water table around the tunnel. The minimum is the elevation of the lowest water table around the tunnel. These values can be estimated readily from topographic maps and piezometers.

ln(2z/r): The rock cover above the tunnel should be known from borings. The tunnel diameter should be known from the design. Since both terms are in a logarithm, the practical range is quite small. For a 2 meter tunnel at 1000 meters deep, ln(2z/r) = 6.9. For a 10 meter tunnel 30 meters deep, ln(2z/r) = 1.8. This gives an extreme range between about 2 and 7.

K: In fractured rock, permeability ranges over many orders of magnitude within a given rock mass. This variability is difficult to predict. The practical minimum for tunnels is around 10-6 cm/s, because even with large heads over long distances, the amount of inflow is very small. The practical maximum, however, can range up to 0.1 cm/s or higher, depending on the rock conditions.

2.3 Heuer method

Heuer (1995, 2005) found that the actual water inflow into tunnel is generally significantly lower than the predicted water inflow using analytical equations and proposed an adjustment factor, on the basis of actual inflow measurements in various tunnels. The adjustment inflow rate is about one-eight of the inflow rate predicted from analytical solutions.

Although this factor is a significant improvement, it cannot be indiscriminately applied to tunnels under a range of conditions and needs to be modified to take into account the effect of depth, hydro mechanical interaction along rock mass discontinuities and other key geological features affecting the tunnel inflow rate.

2.4 Lei method

Lei in 1999, derived analytical expression for hydraulic head, the stream function and the inflow rate of the two-dimensional, steady ground water flow near a horizontal tunnel in a fully saturated , homogeneous, isotropic and semi- infinite porous aquifer for a constant hydraulic head condition at the tunnel perimeter.

$$Q = \left(\frac{2\pi Kh}{\ln\left(\frac{h}{r} + \sqrt{\left(\frac{h}{r}\right)^2 - 1}\right)}\right) \tag{3}$$

Where Q is the groundwater flow, h is the head of water above the tunnel, K is the equivalent permeability and r is the radius of tunnel.(Fig.4)

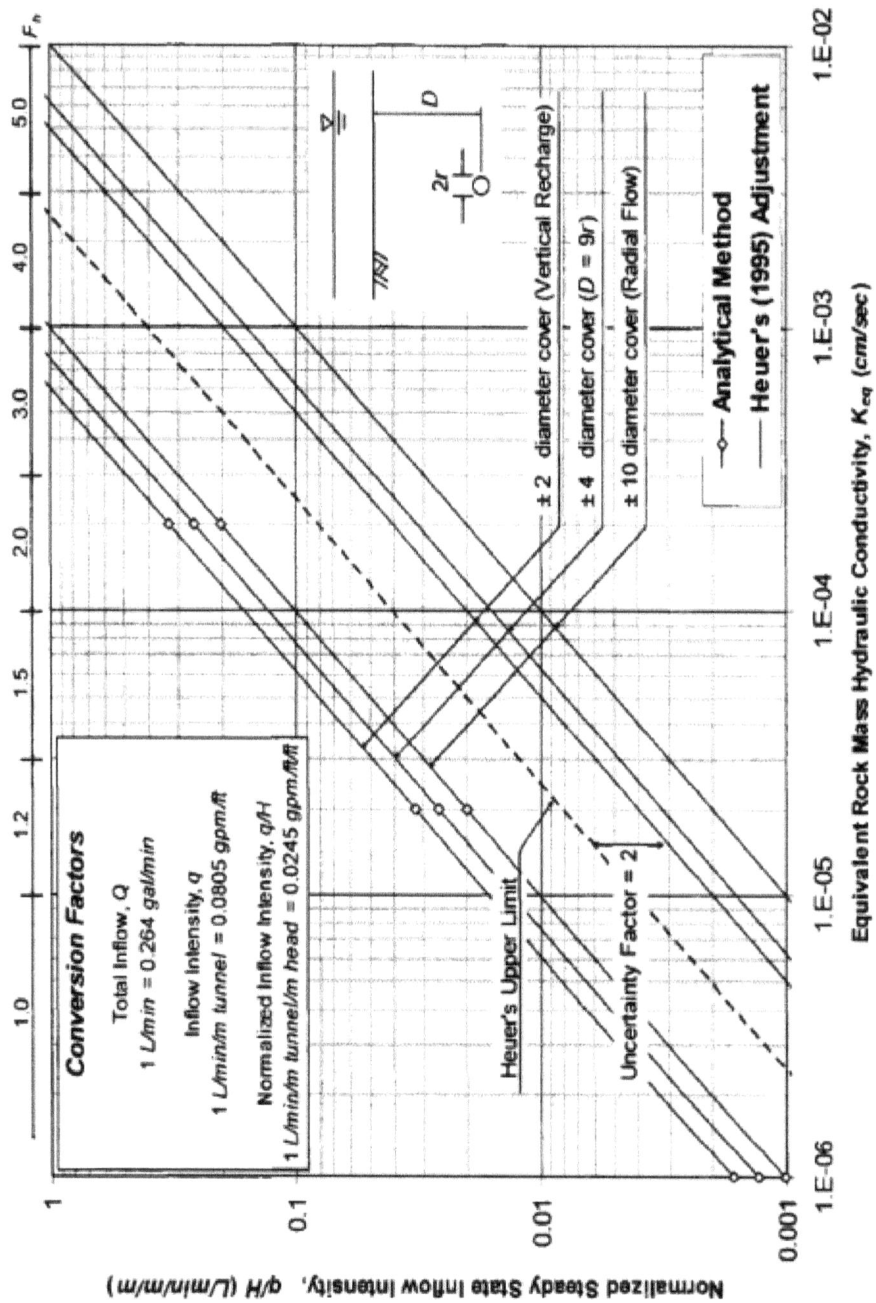

Fig. 3. Relationship between steady state flow and equivalent permeability (Heuer1995).

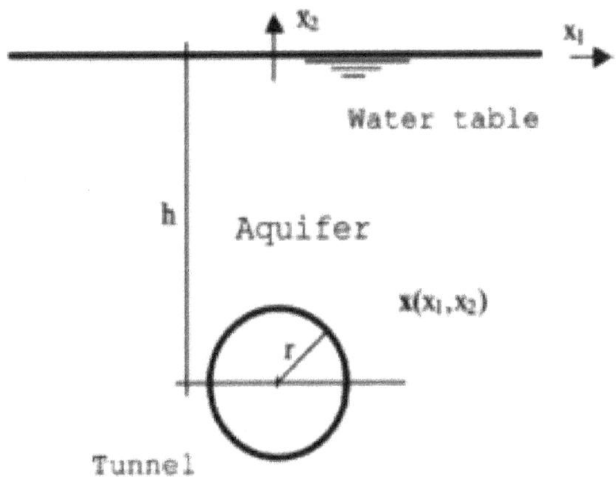

Fig. 4. Circular tunnel in a semi-infinite aquifer with a horizontal water table.

This method deals with an ideal situation. And valid only for cases under the given assumptions. For more complicated scenarios, a numerical model would be more flexible.

2.5 El-Tani method

An accurate solution is derived for the continuous gravity flow caused by a circular tunnel that disrupts the hydrostatic equilibrium of a semi-infinite, uniform, and isotropic aquifer. The flow exhibits various characteristics that may be expressed mathematically, such as the inflow of water, pressure, leakage, and infiltration for recharging. These mathematical expressions provide a comprehensive understanding of the aquifer in a condition of continuous drainage.

$$Q = 2\pi kh\left(\frac{1-3\left(\frac{r}{2h}\right)^2}{\left[1-\left(\frac{r}{2h}\right)^2\right]\ln\frac{2h}{r}-\left(\frac{r}{2h}\right)^2}\right) \quad (4)$$

Where Q is the groundwater flow, h is the head of water above the tunnel, K is the equivalent permeability and r is the radius of tunnel.

It is found that the maximum value of the recharging infiltration does not exceed the hydraulic conductivity allowing stating criteria for recharge intervention to ensure the stability of the aquifer. In addition to the main results, two aspects of the water inflow are treated. These are the necessary modifications that are to be considered in the case of an inclined water table and in the case of a lined tunnel that develops a constant internal pressure. It is also found that under an inclined water table a tunnel may cease to drain on its complete circumferential edge and a limiting condition is stated. Furthermore, the Muskat–Goodman and other water inflow predictions are compared to the exact gravity water inflow.

In this method The gravity flow that isgenerated by a circular tunnel is solved exactly. Other cases need to be obtained in closed forms such as the flow generated by non-circular tunnels in non-homogeneous and bounded aquifers. The integral formulation will probably be useful if it is adequately extended to these cases and to others in the three-dimensional space.

3. TUNNEL SITE RATING FROM GROUNDWATER HAZARD POINT OF VIEW(SGR)

Katibeh and Aalianvari(2009) using the experiments due to ten tunnels in Iran have been proposed a new qualitative and quantitative method for rating the tunnel sites in groundwater hazard point of view, named " Site Groundwater Rating" (SGR). In this method, the tunnel site, according to the preliminary investigations of engineering geological and hydrogeological properties, is categorized into six rates as follow: No Risk, Low Risk, Moderate Risk, Risky, High Risk and Critical. Considered parameters in this method are: joint frequency, joint aperture, karstification, crashed zone, schistosity, head of water above tunnel, soil permeability and annual raining.

In SGR method, after scoring each parameter, the SGR factor of the site is computed and according to this factor the tunnel site is divided into six categories. One of the advantages of this method is helping the engineers and contractors to design more suitable drainage systems and choosing the suitable drilling methods, according to the potential of the groundwater inflow, calculated in SGR.

In general, tunnel site are divided into two parts: tunnels in saturated zones and tunnels in unsaturated zones. Also with respect to the site lithology, tunnel sites are divided into rock sites and soil sites. The equation to compute the SGR factor is:

$$SGR = \left[(S_1 + S_2 + S_3 + S_4) + S_5\right] S_6\ S_7 \qquad (5)$$

where parameters S1 to S7 respecting to the affecting parameters in groundwater inflow, will be described as below.

3.1 Score of frequency and aperture of joints, S1

Massive rocks in tunnels alignment include one or more joint sets and tunnels cut them. Amount of water inflow into tunnels depends on joint frequency and joint aperture, so the representative parameter, S_1, is calculated using Eq.6:

$$S_1 = 25 \times \left(\sum_{i=1}^{n} \frac{\lambda_i\, g e_i^2}{12v} a\right) \qquad (6)$$

where: λi, the joint frequency (1/m), ei, the mean hydraulic joint aperture (m), g, the earth gravity (m/s2), n , the kinematic viscosity of water (m2/s), a, the unit factor (s/m) converting S1 to dimensionless form.

The constant coefficient in Eq.6, 25, is obtained according to the experiments to normalize the parameter S1.

The joint hydraulic aperture, eh, in Eq.7, is different from the joint aperture estimated in surface. Cheng (1994) suggested the following equation to calculate joint hydraulic aperture in depth:

$$e_h = E - \Delta V_j \tag{7}$$

where, E is average joint hydraulic aperture in surface (mm) and ΔV_j is calculated using the following equation:

$$\Delta V_j = \frac{\sigma_n V_m}{K_{ni} V_m + \sigma_n V_m} \tag{8}$$

where, Cheng (1994) suggested σ_n =0.027Z, according to his experiments, in which Z is overburden thickness. Moreover, analyses by Bandis et al. of experimental data indicated that the following relation was appropriate

$$V_m = A + B(JRC) + C(\frac{JCS}{E})^D \tag{9}$$

in which constants A, B, C, and D, according to the Priest's experimental researches are: A= -0.2960, B= -0.0056, C= 2.241, D=-0.2450

JRC is joint roughness coefficient, JCS is joint wall strength and Kni is computed using Eq. 10.

$$K_{ni} = 0.02(\frac{JCS}{E}) + 1.75JRC - 7.15 \tag{10}$$

Finally, S1 is calculated in dimensionless form. If the joint is filled with some materials such as clay, calcite, etc, then ei (Eq. 6) will be equal to zero. However, the caution must be taken for the joints filled with washable materials such as some clay types. Joints with washable materials can be identified in the supplementary site investigation during Lugeon tests.

3.2 Schistosity, S_2

Typically, clay-based rocks undergo schistosity as a result of tectonic events, allowing water to pass through the planes of schist. However, the relevant permeability is significantly lower compared to the other discontinuities. Despite having a poor ability to allow substances to pass through, the S2 parameter in SGR, which represents the presence of schistosity, is expected to fall within the range of 1 to 5, depending on the extent of schistosity.

3.3 Crashed zone, S_3

Crashed zones are the major path of groundwater flow through rock. Crashed zones considerably increase rock permeability; however this increase depends on the rock type. In clay-base rocks such as marl, shale, schist etc, clay minerals fill fractures and discontinuities resulting in considerable decrease in the permeability of crashed zone, but in the other rock types such as limestone, the permeability in crashed zone is very high. Moreover, the groundwater flow rate through crashed zones is related to the rock type and the crashed zone width. Considering rock type and crashed zone width, table 1 shows the equations to calculate S3 in different rocks type in SGR.

Table.1. Method to estimate S3 in crashed zones.

Type of rock	Crashed zone width	S3
Clay base rocks	Czw	2×Log(10Czw× b*)
Other rock type	Czw	100×Log(10Czw**×b)
* b is the unit factor (1/m), **Czw is the crash zone width (m)		

Crashed zones in both saturated and unsaturated zones are suitable paths for groundwater flow, thus, crashed zones are of most importance in SGR.

3.4 Karstification, S_4

Karstification is the geologic process of chemical and mechanical erosion by water on soluble bodies of rock, such as limestone, dolomite, gypsum, or salt, at or near the earth's surface. Karstification is exhibited best on thick, fractured, and pure limestone in a humid environment in which the subsurface and surface are being modified simultaneously. The resulting karst morphology is usually characterized by some types of cavities and a complex subsurface drainage system. So these cavities can conduct groundwater into tunnels. Groundwater inflow into tunnels can be very sudden and so dangerous. According to the degree of karstification, S4 is estimated between 10 to 100.

3.5 Soil permeability, S_5

Parameters S_1 to S_4 are related to rock tunnels but if tunnel is excavated in soil, parameters S_1 to S_4 are automatically equal to zero. In SGR soil permeability is very important factor which is scored in S_5. The permeability of a clay layer can be as low as 10^{-10} m/s, of a weakly permeable layer 10^{-6} m/s and of a highly permeable layer 10^{-2} m/s.

Because of the direct relation between soil permeability and rate of groundwater inflow, in SGR, the score of soil permeability, S_5 is calculated as follow:

$$S_5 = K \times c \tag{11}$$

where, K is the soil permeability (m/day), c is the unit factor (day/m) converting S5 to dimensionless form.

3.6 Water head above tunnel, S_6

Head of water (H) above tunnel is one of the most effective parameters on groundwater inflow into tunnels. The inflow equations such as Muskat–Goodman,

Rat–Schleiss–Lei, Karlsrud and Lombardi indicate that groundwater inflow into tunnel has linear relation with H/Ln(H), so the representative parameter S6, is calculated using Eq.12:

$$S_6 = \frac{H}{Ln(H \times d)} \times d \qquad (12)$$

where, H is water head above tunnel and d is the unit factor (1/m) converting S6 to dimensionless form. When tunnel is excavated above water table S_6 is equal to unit.

3.7 Annual raining, S_7

Just when tunnel is excavated in unsaturated zones, annual raining is effective on groundwater inflow into tunnels. In such case infiltrated water rain can seep into tunnel through fractures and faults. However, in such case groundwater inflow is not permanent like tunneling in saturated zones. Related to the tunnel depth, overburden permeability and length of water channel from discharging area up to tunnel, the time of reaching surface water to tunnel is different. In unsaturated zones quantity and intensity of raining affect the groundwater inflow into tunnels, but here only annual raining is considered. S_7 for unsaturated tunnels is calculated using Eq. 13:

where Py is annual raining (mm).

$$S_7 = \frac{P_y}{5000} \qquad S_7 \leq 1 \qquad (13)$$

Maximum value of S7 is when annual raining is equal to 5000 mm or more. When tunneling in saturated zone, S_7 =1.

3.8 SGR factor

After calculating all the parameters, S_1–S_7, SGR factor of the site is computed by means of Eq.5, then according to the value of SGR and using table 2, the tunnel site category can be found in six cases as: No Danger, Low Danger, Relatively Dangerous, Dangerous, Highly Dangerous, and Critical.

Table 2. SGR rating for groundwater inflow into tunnels.

SGR	Tunnel Rating	Class	Probable conditions for groundwater inflow into tunnel (Lit/s/m)
0-100	No Danger	I	0-0.04
100-300	Low Danger	II	0.04-0.1
300-500	Relatively Dangerous	III	0.1-0.16
500-700	Dangerous	IV	0.16-0.28
700- 1000	Highly Dangerous	V	Q>0.28 Inflow of groundwater and mud from crashed zones is probable.
1000<	Critical	VI	Inflow of groundwater and mud is highly probable.

Experiments due to 10 tunnels show that there are direct correlations between SGR factor and groundwater inflow rate into tunnels. In case of high SGR factor (more than 700), mixture of mud and groundwater is probable to rush into tunnel, endangering persons and equipments.

In the other hand, with attention to SGR factor, groundwater inflow into tunnel can be predicted, which help to plan suitable drainage systems and even to choose the best drilling method.

4. DISCUSSION

This chapter focuses on the methods used to estimate the flow of groundwater into tunnels that are excavated in fractured rock below the water table. This text presents a number of mathematical equations used to determine the amount of groundwater flowing into tunnels. These equations include Goodman (1965), Lohman (1972), Zhang (1993), Heuer (1995), Lei (1999), Karlsrud (2001), Raymer (2001), and El Tani (2003). Additionally, the SGR approach is discussed.

El Tani (2003) compared the results of above mentioned methods with the observed seepage into tunnels, for different values of r/h. Table 3 contains a listing of diverse approximations of the water inflow including Muskat–Goodman, Rat-Schleiss–Lei, Karlsrud, Lombardi and El Tani methods. The relative differences of the diverse formula of table 1 with the exact (observed) water inflow (Q) are shown in Fig. 5 and are computed with:

$$\Delta = \frac{Q_{ap} - Q}{Q}$$

in which, Q_{ap} is a water inflow approximation.

Table 3. Diverse approximations for the water inflow.

Goodman	$Q = 2\pi K \frac{h}{\ln(\frac{2z}{r})}$
Karlsrud	$Q = \frac{2\pi K H_o}{\ln(\frac{2H_o}{r})}$
Heuer	$Q_L = \frac{2\pi K H_o}{\ln(\frac{2z}{r})} \times \frac{1}{8}$
Lee	$Q = 2\pi K \frac{h}{\ln(\frac{h}{r} + \sqrt{(\frac{h}{r})^2 - 1})}$
El Tani	$Q = 2\pi K \frac{1 - 3(\frac{r}{2h})^2}{[1 - (\frac{r}{2h})^2]\ln(\frac{2h}{r} - (\frac{r}{2h})^2}$

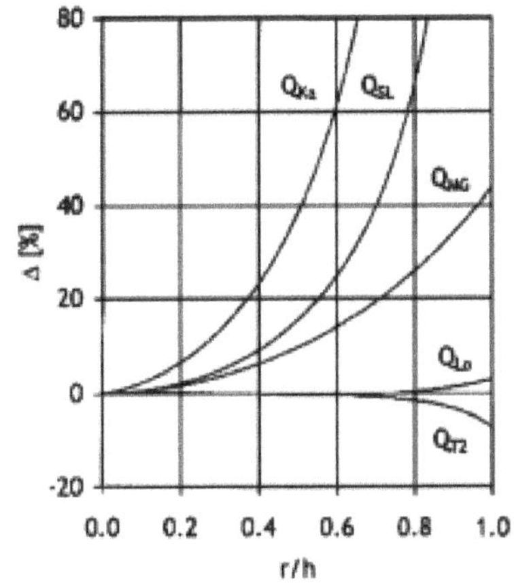

Fig. 5. Relative difference of the diverse approximations in Table 1 with the exact water inflow.

As the Fig. 5 shows, with decreasing r/h, water inflow approximations converge each other and converge to the exact water inflow. For r/h less than 0.3, the relative differences are negligible (El Tani, 2003). For large values of r/h, means tunnels near to the Groundwater table, Lombardi and El Tani approximations are more close to the exact water inflow, while, the relative differences of other equations are considerable in this case (Fig. 5).

A new qualitative and quantitative method for tunnel site rating in groundwater hazard point of view (SGR), was introduced in this chapter. In this method, the tunnel site, according to the preliminary investigations of engineering geological and hydrogeological properties, is categorized into six rates as follow: No Danger, Low Danger, Relatively Dangerous, Dangerous, Highly Dangerous, and Critical. (Katibeh, Aalianvari, 2009). Using SGR factor, groundwater inflow into tunnel can be estimated, which helps to predict suitable drainage system and to choose the best drilling method.

REFERENCES

El Tani, M.; 2003; "Circular tunnel in a semi-infinite aquifer", J. of Tunneling and Underground Space Technology, Vol. 18, PP. 49-55.

Freeze, R.A.; Cherry, J.A. (1979) "Groundwater Englewood", New Jersey, Prentice-Hall Inc, TIC 217571.

Goodman, R., D. Moye, A. Schalkwyk, and I. Javendel.1965. "Groundwater inflow during tunnel driving", Engineering Geology vol.1 , pp 150-162.

Heuer, R.E. 1995. "Estimating rock-tunnel water inflow". Proceeding of the Rapid Excavation and Tunneling Conference, June 18-21.

Katibeh, H., Aalianvari, A.; 2009; “Development of a new method for tunnel site rating from groundwater hazard point of view”, Journal of Applied Sciences, Vol 9, 1496-1502.

Karlsrud, K., 2001. Water control when tunnelling under urban areas in the Olso region. NFF publication No. 12, 4, 27–33, NFF.

Lei, S., 1999. An analytical solution for steady flow into a tunnel. Ground Water 37, 23–26.

Raymer, J.H.,(2003), “Predicting groundwater inflow into hard-rock tunnels : Estimating the high- end of the permeability disterbution”,RETC. Pp. 201-217

05 BANK STORAGE DRAINAGE IN SHALLOW UNCONFINED AQUIFERS

1. INTRODUCTION: AN OVERVIEW

This study focuses on underground drainage systems. For the past 50 years, researchers have been primarily focused on addressing the challenges associated with draining excess water from canals and rivers, both on the surface and underground. Therefore, our study is dedicated to finding suitable solutions for the issues encountered in draining bank storage in shallow unconfined aquifers.

Developing specialized procedures, such as drains and filters, is essential for effectively draining excess water from the canal. It is important to carefully select the most efficient form of drain for this purpose. We conducted a study using a simplified model of a uniform soil with a trench drain on a non-permeable foundation. We have also developed a correlation to identify the optimal placement of the trench drain within the uniform soil. The water level in the trench drains can be accurately determined by considering the head of water in the canal, slope of the canal, and permeability coefficient. This simplifies the seepage analysis to some extent and ensures satisfactory accuracy using the new approach. Ultimately, a comparison was made between the experimental and numerical findings utilizing the [SEEP] software.

2. HORIZONTAL DRAINAGE

2.1 Steady and parallel flow in unconfined aquifers

In this section, we discuss the flow of groundwater to trench drains under steady state conditions. The steady-state theory is based on the assumption that the rate of recharge to the groundwater and evaporation is null.

Figure 1 shows a typical cross-section of a drainage system under this condition.

To describe the flow of groundwater to the trench drains, we have to make the following assumptions:

- Two-dimensional flow. This means that the flow is considered to be identical in any cross-section perpendicular to the drains; this is only true for infinitely long drains;

- Homogeneous and isotropic soils with a permeability *K*. Thus we ignore any spatial variation in the hydraulic conductivity within a soil layer.

Most drainage equations are based on the Dupuit-Forchheimer assumptions (Ritzema, 1994):

Dupuit -Forchheimer equation:

$$\frac{K}{2}\left[\frac{\vartheta^2(h)^2}{\vartheta x^2}+\frac{\vartheta^2(h)^2}{\vartheta y^2}\right]\pm q=0 \qquad (1)$$

Where q = recharge rate per unit area (m/ day /m²) and K = hydraulic conductivity of the soil (m/ day)

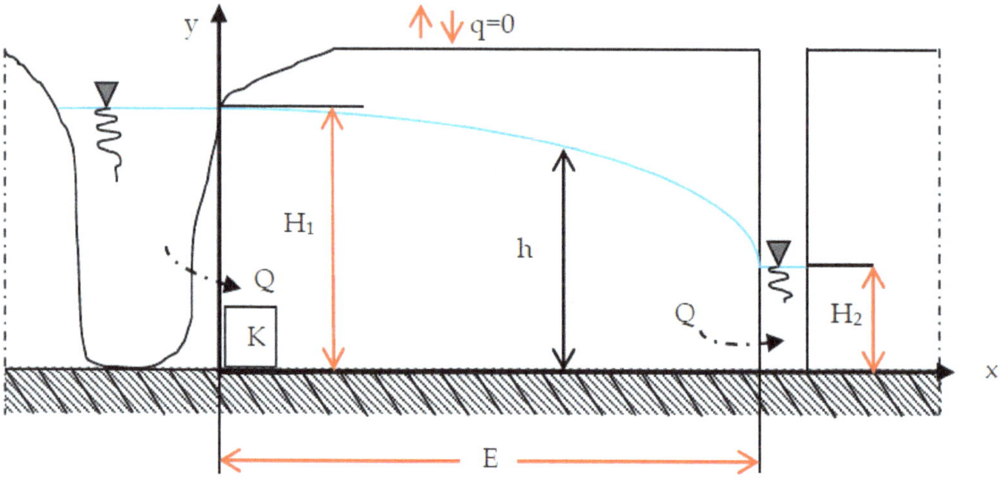

Fig. 1. Cross-sections of trench drains.

when q=0; This equation allows us to reduce the two-dimensional flow to a one-dimensional flow by assuming parallel and horizontal stream lines.

$$\frac{d^2(h)^2}{dx^2}=0 \qquad (2)$$

Thus the solution of this equation is

$$h^2=C_1x+C_2 \qquad (3)$$

The limits conditions of this equation are

for $x=0\rightarrow h=H_1$

for $x=E\rightarrow h=H_2$

H_1 = elevation of the water level in the canal (m)

H_2 = elevation of the water level in the trench drain (m) E = drain spacing (m)

$$C_1=\frac{H_1^2-H_2^2}{E},\quad C_2=H_1^2 \qquad (4)$$

$$h^2 = H_1^2 - (H_1^2 - H_2^2)\frac{x}{E} \quad (5)$$

2.2 Darcy's law

The movement of water through granular materials was first investigated by Darcy in 1856 when he became interested in the flow characteristics of sand filter beds. In his experiments he discovered the law governing the flow of homogeneous fluids through porous media.

Darcy's Equation can be applied to describe the flow of groundwater (Q) through a vertical plane (h) at a distance (x) from the ditch

$$Q = -V.S = -K.I.h.1 = -Kh\frac{\partial h}{\partial x} = -\frac{K}{2}\frac{\partial(h^2)}{\partial x} \quad (6)$$

which can also be written as;

$$Q = \frac{K(H_1^2 - H_2^2)}{2E} \quad (7)$$

if $\pm q \neq 0$

$$\frac{K}{2}\frac{d^2(h^2)}{dx^2} + q = 0 \quad (8)$$

The solution of this equation is

$$h^2 = -\frac{q}{K}x^2 + C_1 x + C_2 \quad (9)$$

The limits conditions of this equation are

for x=0 → h = H_1

for x=E → h = H_2

$$h^2 = H_1^2 - (H_2^2 - H_1^2)\frac{x}{E} + \frac{q}{K}x(E - x) \quad (10)$$

equations (6) and (10), can be written:

$$Q = \frac{K(H_1^2 - H_2^2)}{2E} + q\left(x - \frac{E}{2}\right) \quad (11)$$

With this equation, we can determine the limit flow for Q=0 (Fig. 2)

at x=0 → = Q^0

x=E → = Q_0

1. Example problem:

A trench drain of 500m effective length is sunk into an aquifer of permeability K=2,0 m/d at 200m from the canal when H_1 =5,0m is the elevation of the water level in the canal, H_2 =0 is the elevation of the water level in the trench drain and the rate of recharge per unit surface area is q=0,01m/d (Fig. 3).

Calculate the flow of groundwater Q= ?

Solution:

We are given:

H_1=5,0m; H_2=0; B=500m; K=2,0m/d; q=0,01m/d; E=200m.

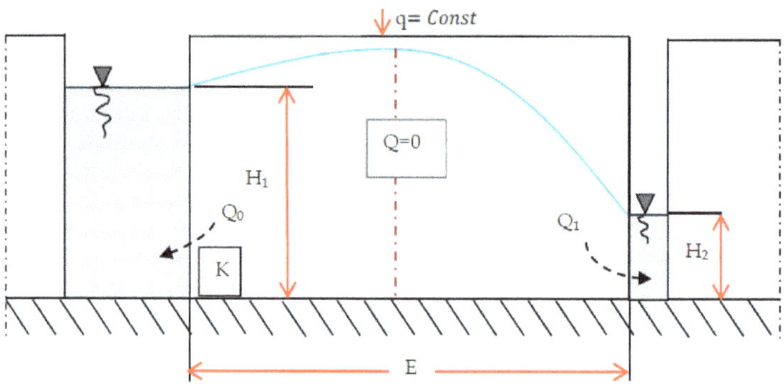

Fig. 2. Cross-sections of open field drains, showing a curved watertable.

The flow of groundwater (Q) through a vertical plane (*h*) at a distance (x) from the ditch is given by this equation

$$Q = \frac{K(H_1^2 - H_2^2)}{2E} + q\left(x - \frac{E}{2}\right)$$

Numerical application:

a. $x=0; \rightarrow Q_0 = \frac{2(5^2-0)}{2.200} + 0{,}01\left(0 - \frac{200}{2}\right) = -0{,}875\frac{m^3}{d.ml}$

b. $x=E; \rightarrow Q_0 = \frac{50}{400} + 0.01\left(200 - \frac{200}{2}\right) = 1{,}125\frac{m^3}{d.ml}$

For the whole length B, we have:

$$\sum Q = Q.B: \sum Q_0 = -0{,}875.500 = -437{,}5\frac{m^3}{d} = -5{,}06\ l/s$$

$$\sum Q_E = 1{,}125.500 = 562{,}5\frac{m^3}{d} = 6{,}51\ l/s$$

2[sec] variant: when q=0

The discharge can be expressed by the following formula:

$$Q = \frac{K(H_1^2 - H_2^2)}{2E}$$

Numerical application:

$$Q = \frac{2(5^2-0)}{2.200} = 0{,}125\frac{m^3}{d.ml'};$$

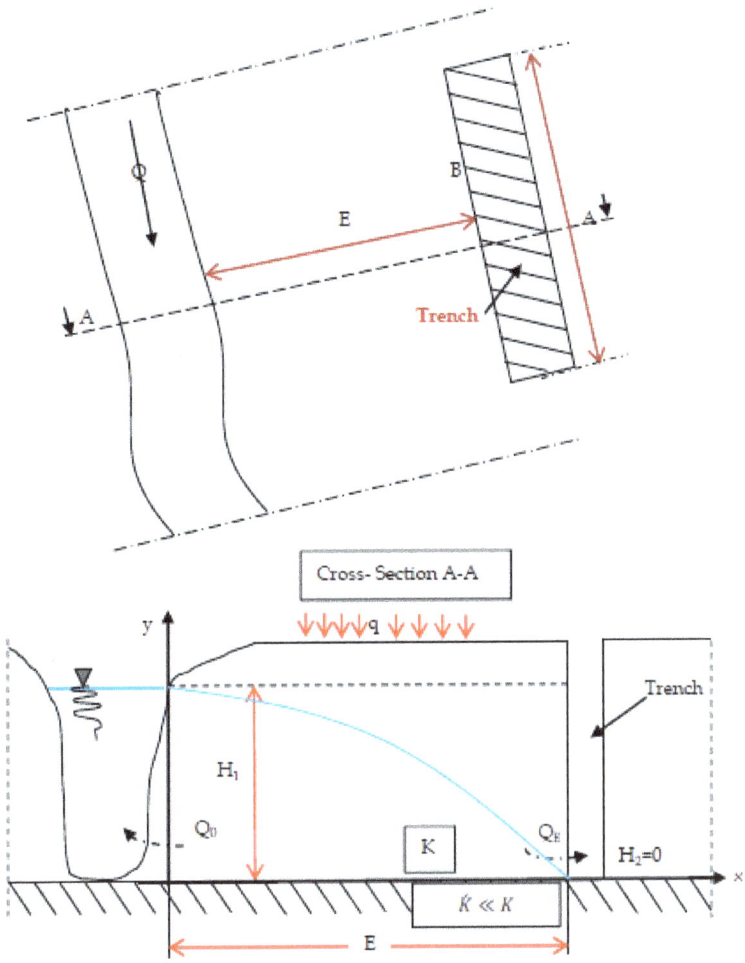

Fig. 3. Cross-sections of open field drains, showing a curved watertable under recharge from rainfall, or irrigation.

$$\sum Q = Q.B: \sum Q = 0{,}125.500 = 62.5\frac{m^3}{d} = 0{,}72\ l/s$$

2. Example problem:

A trench drain of 500m effective length is sunk into an aquifer of permeability K at 200m from the canal, when H_1 =5,0m is the elevation of the water level in the canal, H_2 =0 is the elevation of the water level in the trench drain, the rate of recharge per unit surface area is $\Sigma Q = 100\frac{m^3}{d}$

Calculate the permeability of the aquifer ***K*** = ?

Solution:

We are given:

H_1=5,0m ; H_2=0 ; B=500m ; q=0,01m/d ; E=200m; Σ Q=Q.B =100 $\frac{m^3}{d}$.

the permeability of the aquifer **K** is:

$$K = \frac{2QE}{(H_1^2 - H_2^2)} = \frac{2.0,200.200}{(5^2-0)} = \frac{80}{25} = 3,2\frac{m^3}{d};$$

the flow of groundwater Q:

$$Q = \frac{\Sigma Q}{B} = \frac{100}{500} = 0,2\frac{m^3}{d.ml}$$

2.3 Grapho-Analytical Method for parallel drainage

The pumping rate is estimated by the formula of Chapman (Leonards, 1968)

$$q = \frac{K}{r}(H - 0,27h_0)(H^2 - h_0^2) \quad (12)$$

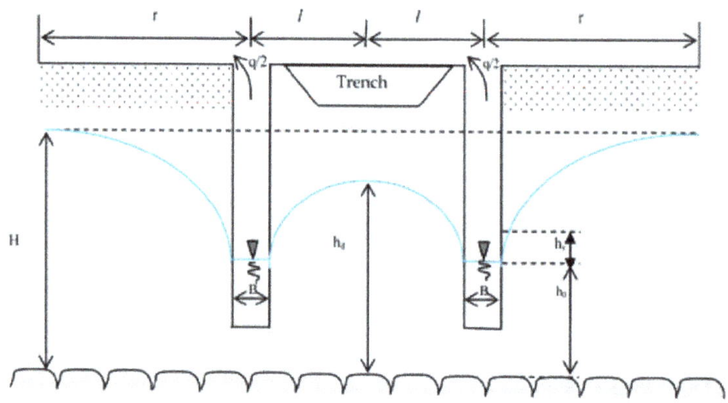

Fig. 4. Flow of groundwater to subsurface drains in shallow unconfined aquifers.

The following equation can be used to determine the elevation of the watertable midway between the drains:

$$h_d = h_0\left[\frac{C_1C_2}{r}(H - h_0) + 1\right] \quad (13)$$

C_1 and C_2 are estimated by means of the curves of the flowing figure.

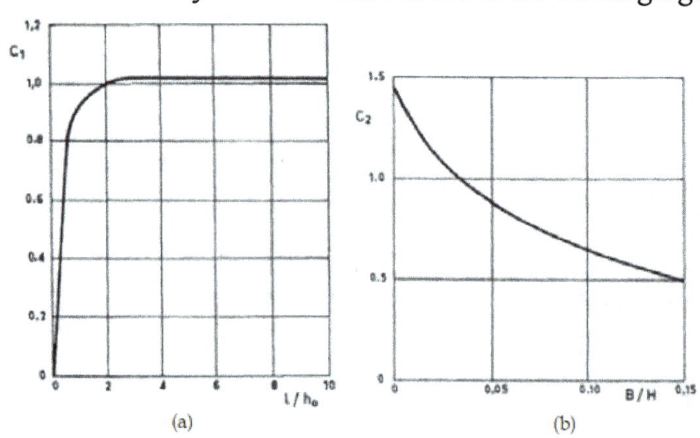

Fig. 5. Values of C_1 (a) and C_2 (b) (from Cassan, M. (1994)).

3. Example problem:

Calculate the pumping rate q and the elevation of the watertable midway between the drains hd =?

We have:

H=8m ; h_0=5,5m ; 2*l*=20m ; K=9.0m/d ; B=0,8m ; r=50m.

Solution:

the pumping rate is:

$$q = \frac{K}{r}(H - 0{,}27h_0)(H^2 - h_0^2)$$

$$q = \frac{9}{50}(8 - 0{,}27.5{,}5)(8^2 - 5{,}5^2) = 4{,}95\frac{m^3}{d.ml}$$

We can determinate C_1 and C_2 from the curves of the fig.5, we have:

$$\frac{a}{h_0} = \frac{10}{8} = 1{,}25 \rightarrow C_1 = 0{,}93$$

$$\frac{B}{H} = \frac{0{,}8}{8} = 0{,}1 \rightarrow C_2 = 0{,}6$$

And the elevation of the watertable midway between the drains:

$$h_d = h_0\left[\frac{C_1C_2}{r}(H - h_0) + 1\right]$$

$$h_d = 5{,}5\left[\frac{0{,}93.0{,}6}{50}(8 - 5{,}5) + 1\right] = 5{,}65m$$

The drawdown of water level is:

$$H - h_d = 8 - 5{,}65 = 2{,}35m$$

3. SINGLE HORIZONTAL DRAINAGE IN UNCONFINED AQUIFERS

3.1 Perfect drainage

The pumping rate is estimated by the Dupuit formula (Cassan, 1994):

$$H_0^2 - H_p^2 = \frac{Q}{\pi K} ln\frac{r}{r_p} \qquad (14)$$

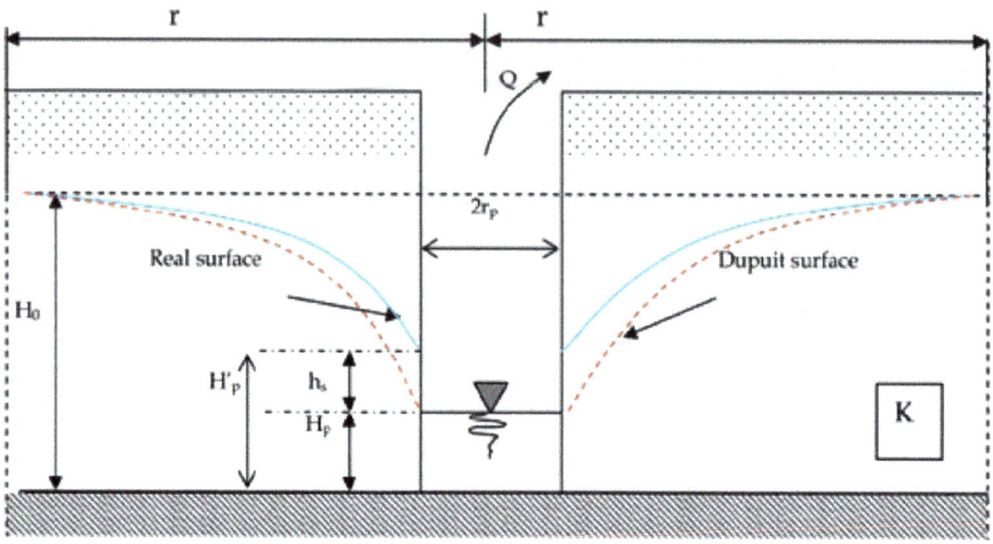

Fig. 6. The drawdown of water level in free nappe.

We can estimate the height of seepage hs from the abacus of schneebeli (Fig.7), if the permeability of the aquifer ***K*** and the pumping rate Q are known.

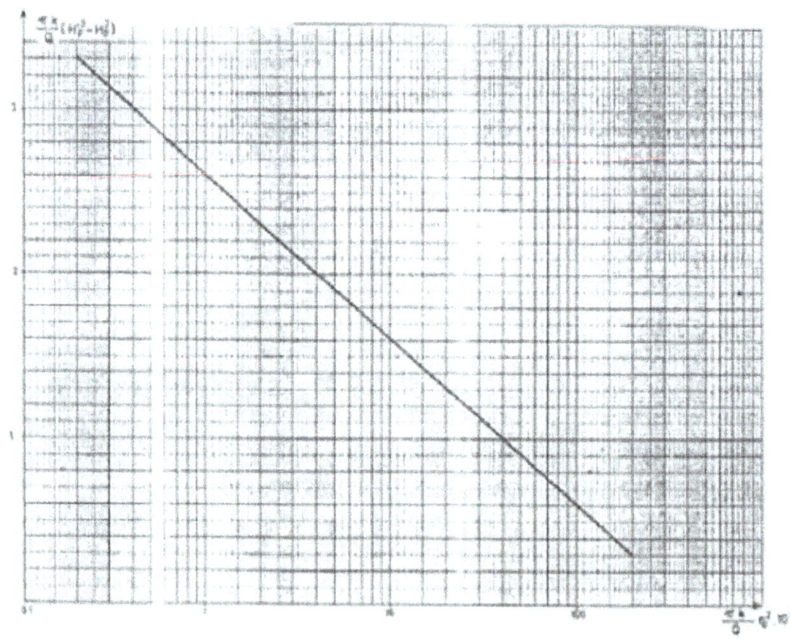

Fig. 7. Height of seepage given by the abacus of schneebeli (from Cassan, M. (1994)).

3.2 Imperfect drainage

The pumping rate is estimated by the flowing formula:

$$q = \frac{K}{r}(H - 0{,}27h_0)(H^2 - h_0^2) \tag{15}$$

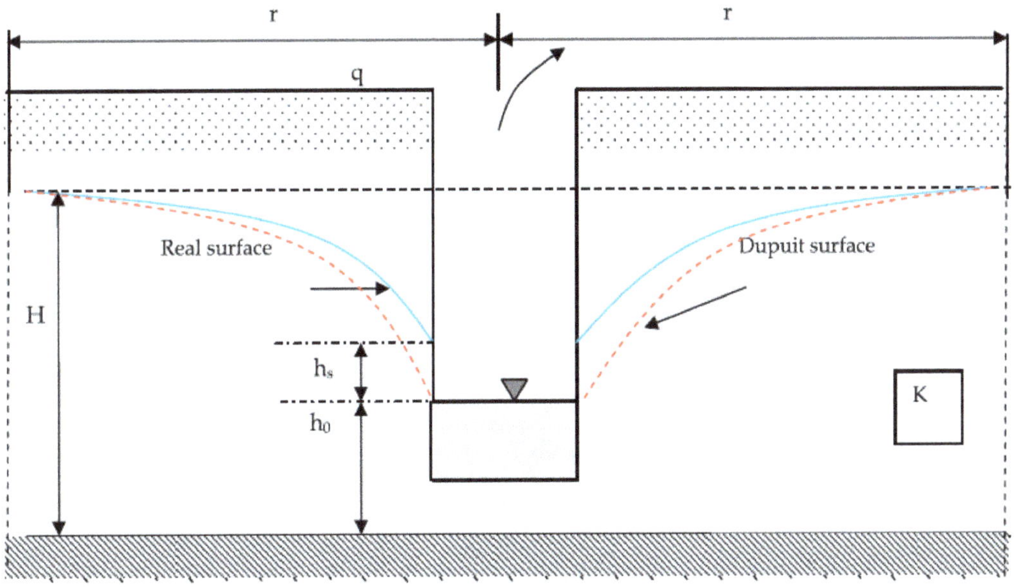

Fig. 8. Imperfect drainage in free nappe.

4. SINGLE HORIZONTAL DRAINAGE IN SEMI-CAPTIVE AQUIFERS

The pumping rate is estimated by the flowing formula (Leonards, 1968):

$$q = \frac{2KD(H - h_0)}{l + \lambda D} \tag{16}$$

where: λ = coefficient given by the curve of the figure 10.

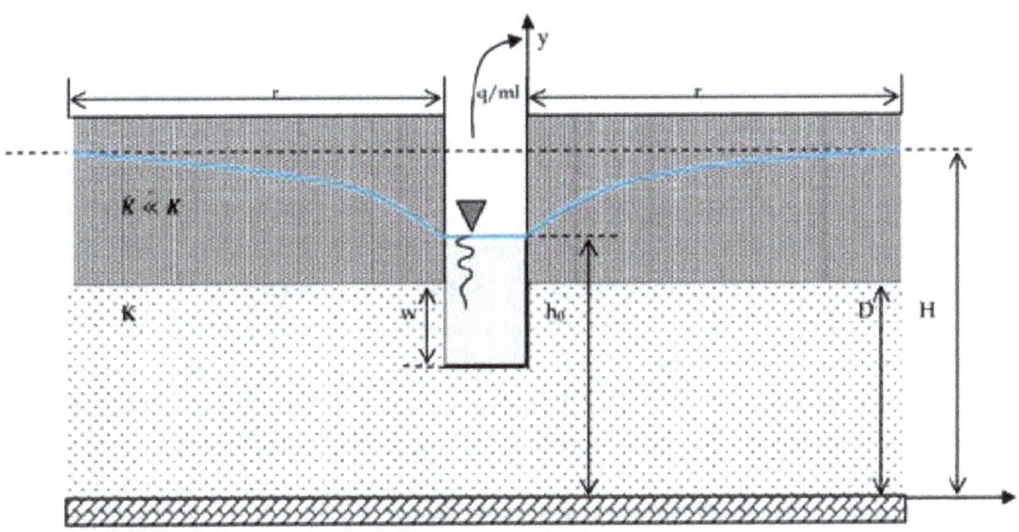

Fig. 9. Single horizontal drainage in semi-Captive aquifers.

where r =radius of influence

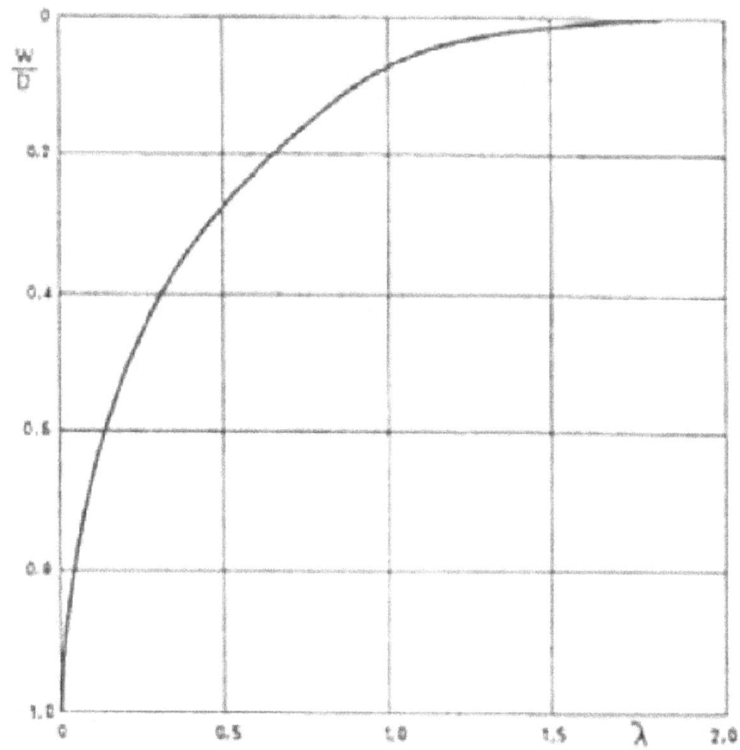

Fig. 10. Values of λ (from Cassan, M. (1994)).

5. DEPLETION OF WATER STORED IN THE BANKS OF SHALLOW, UNCONFINED AQUIFERS

5.1 Shallow unconfined aquifers

Unconfined aquifers are characterized by the existence of a water table that is not restricted by any confining layers. As a result, groundwater in unconfined aquifers has the ability to flow in any direction, including horizontally, vertically, or at angles between these two directions. Shallow unconfined aquifers are characterized by the presence of a relatively close-to-the-surface impermeable layer, often located between 0.5 to 2 meters below the soil surface. The movement of groundwater towards subsurface drains located above a shallow layer that does not allow water to pass through is primarily in a horizontal direction and largely takes place above the level of the drains (see Figure 11). For shallow unconfined aquifers, it is typically satisfactory to determine the horizontal hydraulic conductivity of the soil above the drain level, also known as Ka. Water replenishment to a shallow aquifer is solely dependent on the percolation of rain or irrigation water. There is no upward movement of groundwater or any natural drainage. According to Oosterbaan and Nijland (1994), the horizontal flow in a shallow aquifer is typically ignored when there are no subsurface drains due to the low transmissivity of the aquifer.

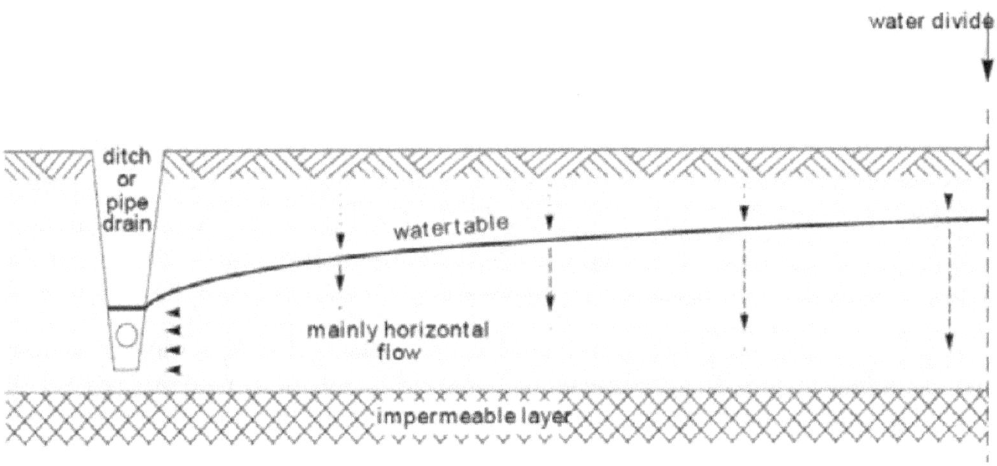

Fig. 11. Flow of groundwater to subsurface drains in shallow unconfined aquifers (from Oosterbaan, R.J. & Nijland, H.J. (1994)).

5.2 Methods for solving unconfined drainage

5.2.1 Introduction

Various methodologies have been employed to address seepage issues in porous materials. Literature has presented graphical, analytical, experimental, and numerical approaches to ascertain the hydraulic characteristics of seepage in uniform soil.

The governing steady flow, given by the Laplace equation (Equation 17), is derived by applying the law of mass conservation with Darcy's Law. This can be solved by constructing a flow net using graphical methods. A flow net is a system of interconnected curves known as streamlines and equipotential lines. A streamline is a theoretical line that delineates the trajectory of a groundwater particle as it moves throughout an aquifer. In an aquifer that exhibits isotropy, the streamlines are oriented at a right angle to the equipotential lines.

The Laplace equation for isotropic soil is:

$$\nabla^2 h = \frac{\partial^2 h}{\partial x^2} + \frac{\partial^2 h}{\partial y^2} = 0 \tag{17}$$

Where h is the total fluid head.

5.2.2 Analytical methods

Flow problems may be solved by analytical, experimental means or a combination of the two. Some analytical solutions are available for seepage flow problems. In a few cases, analytical expressions can be obtained by direct integration of the appropriate differential equation. Harr (1962) explains the use of transformations and mapping to transfer the geometry of a seepage problem from one complex plane to another. Conformal transformation techniques (conformal mapping) and velocity hodograph can also analyze a range of problems (Abd El Razek , & Nasr, 1993).

Pavlovsky (1936, 1956) developed an approximate method which allows the piecing together of flow net fragments to develop a flow net for the total seepage problem. This method, termed the method of fragments, allows rather complicated seepage problems to be resolved by breaking them into parts, analyzing flow patterns for each, and reassembling the parts to provide an overall solution.

In 1931, Kozeny (Abd El Razek , & Nasr, 1993) studied the problem of seepage through an earth dam on an impervious base with a parabolic upstream face and a horizontal underdrain. Using conformal mapping, he obtained a parabola for the free surface.

5.2.3 Numerical methods

The flow through porous media occurs usually in domain shapes that make the possibility of obtaining an analytical solution highly improbable. The nonhomogenity of the soil, the complexity of the boundaries and boundary conditions make the situation even more difficult (Hayder Hassan Al.Jairry, 2010).

The numerical solution of boundary value problems is a very old subject. In the past, the difficulty lied in the huge number of the obtained simultaneous algebraic equations, which have to be solved to reach the solution. The development of strong numerical methods can be applied to almost all types of problems of flow through porous media. These lead to the production of high quality easy to use software that can solve complicated problems and present the results in most convenient forms.

Computer models are used to make acceptable approximations for the Laplace equation in complex flow conditions.

The two primary methods of numerical solution are finite difference and finite element. Both methods can be used in one-, two-, or three-dimensional modeling.

1. The finite difference method solves the Laplace equations by approximating them with a set of linear algebraic equations. The flow region is divided into a discrete rectangular grid with nodal points which are assigned values of head (known head values along fixed head boundaries or points, estimated heads for nodal points that do not have initially known head values). Using Darcy's law and the assumption that the head at a given node is the average of the surrounding nodes, a set of N linear algebraic equations with N unknown values of head are developed (N equals number of nodes). Normally, N is large and relaxation methods involving iterations and the use of a computer must be applied.
2. The finite element method is a second way of numerical solution.

This method is also based on grid pattern (not necessarily rectangular) which divides the flow region into discrete elements and provides N equations with N unknowns. Material properties, such as permeability, are specified for each element and boundary conditions (heads and flow rates) are set. A system of equations is solved to compute heads at nodes and flows in the elements.

The finite element has several advantages over the finite difference method for more complex seepage problems. These include (Radhakrishnan, 1978):

- Complex geometry including sloping layers of material can be easily accommodated.
- By varying the size of elements, zones where seepage gradients or velocity are high can be accurately modeled.
- Pockets of material in a layer can be modeled.

5.2.4 Graphical method

The graphical method can be used to solve a wider class of problems than analytical method. Its advantage is highly remarkable in case of potential flow through domains with irregular boundaries. Other types of flow such as through two dimensional anisotropic homogeneous soil or multilayers can also be treated graphically.

Flow nets are one of the most useful and accepted methods for solution of Laplace's equation (Casagrande, 1937). If boundary conditions and geometry of a flow region are known and can be displayed two dimensionally, a flow net can provide a strong visual sense of what is happening (pressures and flow quantities) in the flow region.

To draw the flow net, the following conditions are to be satisfied as follows (Harry, 1989);

1. All impermeable boundaries are streamlines,
2. All water bodies are lines of constant head (equipotential lines),
3. Streamlines meet equipotential lines at right angles,
4. Streamlines do not intersect one another, and
5. Equipotential lines do not intersect one another.

Using the previous conditions, it is recommended to start the flow net construction by drawing the streamlines taking the advantage of the existing impermeable surfaces in the domain of flow and making sure that they meet water bodies at right angles. However, the graphical method requires a lot of experience and many trial and error works before reaching a solution with reasonable accuracy.

5.2.5 Experimental model

Experimental methods are considered useful for simulating the flow of water by models in laboratory. There are two types of models, electrical model, which analogous the flow of water by a flow of current and physical models such as sandbox model and viscous flow model (Hele-Shaw model). These methods have some inconvenient such as the complicated construction and operation. In Hele-Shaw model, the viscosity of the fluid varies with temperature, and sandbox model suffers from the difficulty of representing the correct permeability of the soil and because of the difficulties caused by capillarity. In this study we have used a sandbox model.

5.2.5.1 Sand-box model

Sand models which may use prototype materials can provide information about flow paths and head at particular points in the aquifer. We have constructed a scale model (generally sand) of the prototype in a sandbox equipped from perforated front and which allow the passage of water. When steady-state flow is reached, the flow can be traced by dye injection at various points along the upstream boundary close to the transparent wall to form the traces of streamlines, and heads determined by small piezometers (El-Masry, 1995; Khalaf Allah, 2005).

A small-scale model was built; which is geometrically similar to the real system. This model represents a homogeneous soil (bank storage in shallow unconfined aquifers) with a trench drain on an impervious foundation. Sand has been used as a permeable medium for the body of the soil, provided that its permeability is such as the flow remains laminar and that there is not any effect of distortion by capillarity (the grains should not be lower than 0,7 mm) (Mallet & Pacouant, 1951) and of gravel for the trench drain. The piezometric prickings laid out on the two zones with dimensions of the tank make it possible to know the actual values of the head of water along the trajectory of flow and highlight the burden-sharing of water in the seepages (Bear, 1972; Casagrande, 1973; Harr, 1962).

6. REDUCED MODEL OF BANK STORAGE IN SHALLOW UNCONFINED AQUIFERS

6.1 Determination of the material characteristics

The characteristics of materials used in this model have been determined (sand for the soil and gravel for the drain) such as the vertical and horizontal permeability

6.2 Vertical permeability

It is given according to the Darcy' law:

$$Q = K_v . SI \tag{18}$$

where Q = quantity of discharge; S = cross-sectional area of flow; I = hydraulic gradient; Kv

= coefficient of vertical permeability.

We obtain, *Kv*= 4.9(m/day) = 5, 67.10-5 (m/s),

6.3 Horizontal permeability

This permeability is given according to the formula of Dupuit:

$$Q = \frac{K_h(H_1^2 - H_2^2)}{2L} \quad (19)$$

where Q = quantity of discharge; H1 = the head of water upstream; H_2 = the head of water downstream; L = the length of the sample; b = the width of the sample; K_h = coefficient of horizontal permeability. We obtain K_h = 43.2 (m/day) = 5.10-4(m/s).

7. TRENCH DRAIN

7.1 Position of the trench drain

The best position of a trench drain in a homogeneous soil (fig.12.) can be determined according to the maximum head of water in the canal, the slope of the canal and the critical hydraulic gradient of the soil, for the case of fine sand, Icr=0,38 (Volkov, 1986). Assuming that the curve of saturation (phreatic line) has a linear shape, we proposed the following relationship (Djehiche, 1993, Djehiche et al., 2006, Djehiche & Kotchev, 2008).

$$L_d = nH_m + \frac{H_m - H_d}{I_{cr}} \quad (20)$$

where H_m = maximum head of water in the canal; h_d = imposed head of water in the trench drain; I_{cr} = critical gradient of material used; n = slope of the canal; L_d = position of the trench drain.

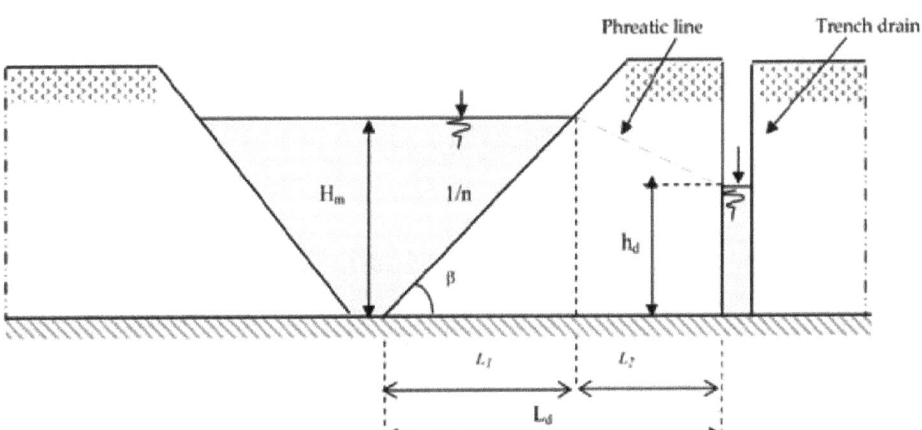

Fig. 12. The position of the trench drains.

7.2 Water level in the trench drain

The water level in the trench drain can be determined for a given head of water in the canal by using the two following relationships (Eq.24 and Eq.25). For the first relationship (Eq.24), we assume that the curve of saturation (phreatic line) has a linear shape, and the critical hydraulic gradient material of the soil is under the following condition:

$$I_{cr} \leq I \leq I_{adm} \qquad \text{and} \qquad I = \frac{0.82H}{L_d - nH}$$

I_{adm} - allowed infiltration gradient (Volkov, 1986).

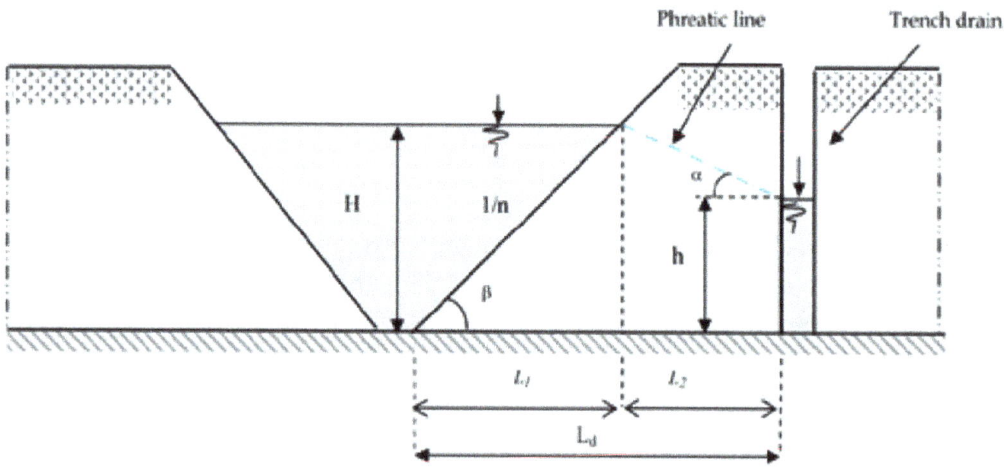

Fig. 13. the water level in the trench drain.

$$\tan\beta = \frac{H}{L_1} = \frac{1}{n} \Rightarrow L_1 = nH \tag{21}$$

$$\tan\alpha = \frac{H-h}{L_2} = I_{cr} \Rightarrow L_2 = \frac{H-h}{I_{cr}} \tag{22}$$

$$L_d = L_1 + L_2 = nH + \frac{H-h}{I_{cr}} \tag{23}$$

The first relationship is (Djehiche et al., 2006).

$$h = H - (L_d - nH)I_{cr} \tag{24}$$

where h = water level in the trench drain[cm]; H = the head of water in the canal [cm]; I_{cr} =

critical gradient of material used; n = slope of the canal; L_d = position of the trench drain[cm]. A comparison between experimental, numerical and the last relationship results is presented in fig. 14.

7.2.1 Discussion

Good agreement can be observed between the curves obtained by the empirical formula (24), the experimental data and the numerical results obtained with "SEEP" (Duncan, 1970). However, if we have $I < I_{cr}$ i.e. for a weak head of water the real gradient is lower than the critical gradient and we cannot use this relation.

The variation of the water level in the drain (h) according to the head of water (H) given by the three procedures of calculation are well adjusted except for the case n=2.5 where there is a small shift between the numerical results and the two others.

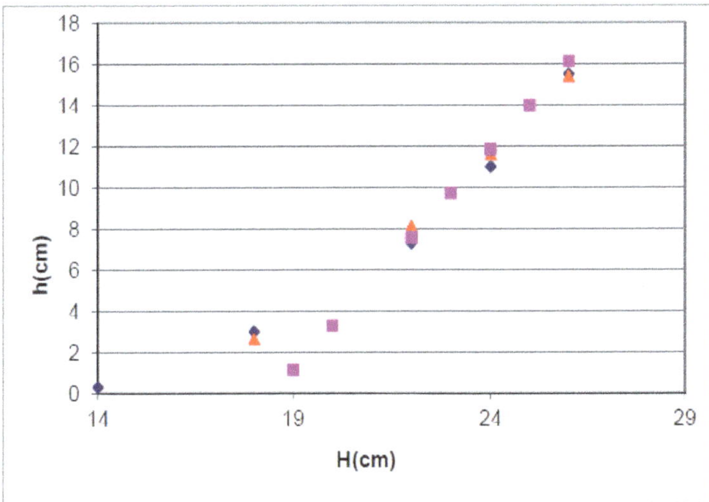

(a) n=3,0

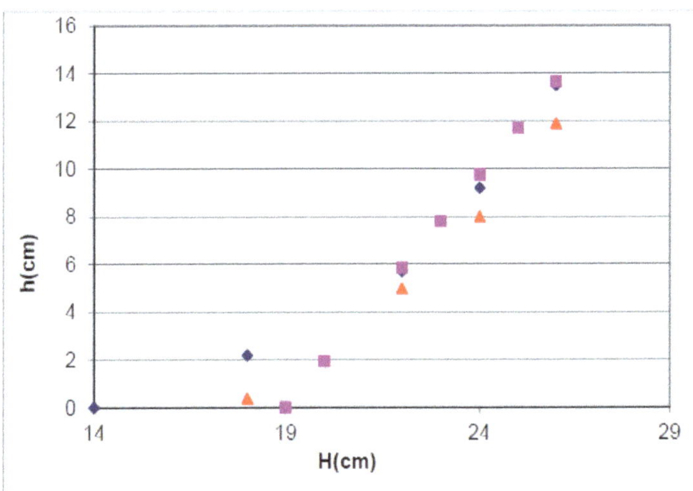

(b) n=2,5

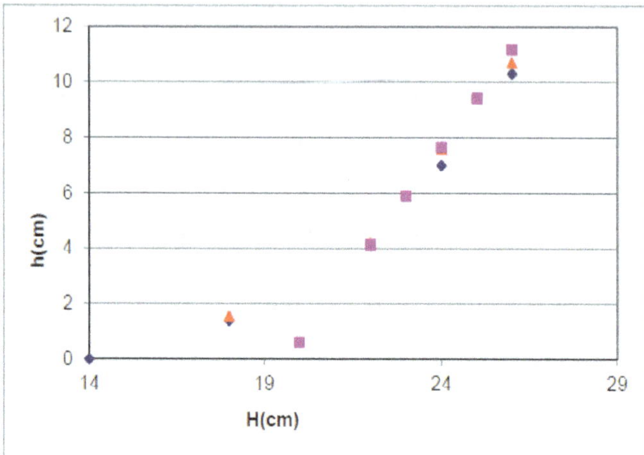

(c) n=2,0

Fig. 14. The lozenge represent the experimental results of the tests, the square represents the results obtained with equation (24), and the triangle represents the numerical results.

- The second relationship under the following condition:

$$5 \leq \frac{K_h}{K_v} \leq 30$$

$$h = e^{\left[\frac{0.18791H\left(\frac{K_h}{K_v}\right)^{0.3}}{\sqrt{n}} - \frac{3.3181\sqrt{\frac{K_h}{K_v}}}{n^{1.2}}\right]} \tag{25}$$

where h = water level in the trench drain[cm]; H = the head of water in the canal [cm]; K_h = coefficient of horizontal permeability; K_v = coefficient of vertical permeability; n = slope of the canal.

The comparison of results of this relationship with the results of SEEP (Duncan, 1970) "software" and the experimental results is represented in fig. 15.

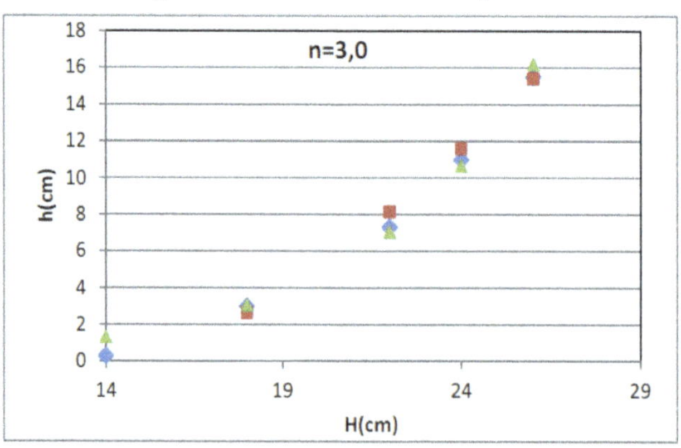

(a)

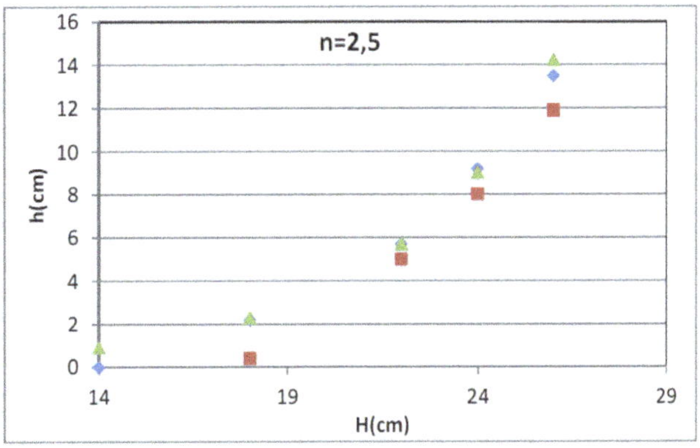

(b)

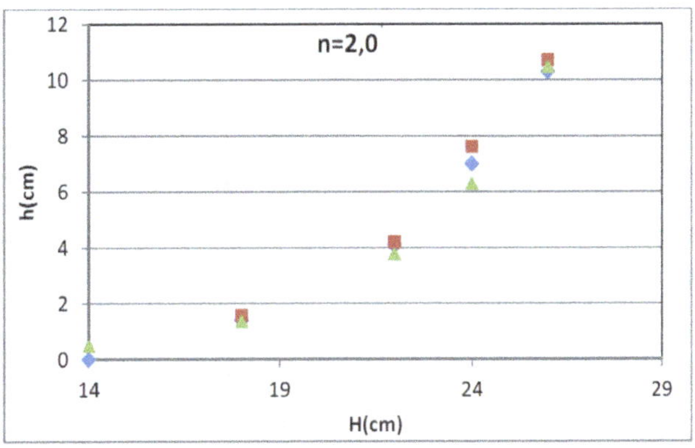

(c)

Fig. 15. The lozenge represents the experimental results of the tests, the triangle represents the results obtained with equation (25), and the square represents the numerical results: (a) n=3,0; (b) n=2,5; (c) n=2,0.

We observe a good agreement between the curves obtained by the three procedures the empirical formula (25), the experimental data and the numerical results obtained with "SEEP" (Duncan, 1970).

8. CONCISE SUMMARY

The models created in this study allow us to determine the optimal placement of trench drains for draining bank storage in shallow unconfined aquifers. The water level in the trench drains can be accurately determined by considering the head of water in the canal, slope of the canal, and the permeability coefficient. This simplifies the seepage analysis to some extent and ensures satisfactory accuracy using the new approach. Furthermore, these equations are easily applicable and may be utilized for the purpose of designing and examining any uniform soil with trench drains in the drainage of bank storage in shallow unconfined aquifers. These highly promising findings allow us to extend the application of these results to the domain of a network of drains in a drainage system. Additionally, we can now examine predicting the seepage through earth dams and the occurrence of piping phenomena, which are frequently the cause of numerous catastrophic incidents.

REFERENCES

Abd El Razek M., & Nasr R. I., (1993). Evaluation of the Minimum Length of Filter in Earth Dams, Journal *of Alexandria Engineering*, Vol. 32, No. 4, October 1993, ISSN 1110-0168.

Bear, J.; (1972). *Dynamics of Fluids in Porous Media*, Elsevier, ISBN 0-444-00114-X, New York Casagrande,A. (1973). *Seepage control in earth dams.* J. Wiley & Sone.

Chu , J.; Bo, M. W.; & Choa, V. (2004). Practical considerations for using vertical drains in soil improvement projects, *Geotextiles and Geomembranes* , N° 22: pp 101–117, ISSN 0266-1144

Cassan, M. (1994). *Aide-Memoire D'hydraulique Souterraine.* Presses de l'école nationale des Ponts et Chaussées (ENPC) - 2ème Édition, France. ISBN-10: 2-85978-208-7

Djehiche, A. (1993). *Comportement des barrages en terre avec cheminée filtrante sous l'action de l'infiltration.* Thèse Magister, Algeria: USTO, pp 121 (in Arabic).

Djehiche, A.; Derriche, Z. & Kotchev, K. (2006). *Control of water head in the vertical drain.* Proceedings of Dams and Reservoirs, Societies and Environment in the 21st Century, Vol.1, pp 913-916, ISBN-10: 0-415-40423-1.

Djehiche, A.; & Kotchev, K.; (2008). *Control of seepage in earth dams with a vertical drain.* Chinese Journal of Geotechnical Engineering, pp 1657-1660, Vol.30, Nov (2008), ISSN *1000-4548.*

Duncan, J. M. (1970). SEEP, A computer for seepage with a free surface or confined steady flow. University of California: Berkeley.

Dunglas, J. & Loudiere, D. (1973). Nouvelle conception des drains dans les barrages en terre homogènes de petite et moyenne dimensions. *La Houille Blanche*, 5(6): pp. 461-465, ISSN 0018-6368.

El-Masry A. A., (1995). Unconfined Seepage through Earth Dams with Chimney Drain by Boundary Element Method", Proceeding of First Engineering conference, Mansoura, March 1995, Vol. III. pp. 357-365.

Harr, M. E.; (1962). *Groundwater and Seepage*, McGraw-Hill Book Company, ISBN 0-486- 66881-9, New York.

Harry R. Cedergren, (1989). *Seepage, drainage, and flow nets.* John Wiley & Sons, Inc., New York, NY.

Hayder Hassan Al.Jairry (2010) 2D-Flow Analysis Through Zoned Earth Dam Using Finite Element Approach, *Eng. & Tech. Journal* ,Vol.28, No.21,

Kahlown M. A. & Khan A.D. (2004). *Tile drainage manual.* Khyaban-e-Johar, H-8/1, Islamabad – Pakistan ISBN 969-8469-13-3

Khalaf Allah S., (2005). *Seepage through Earth dams with Filters,* M. Sc thesis, Dept. of Irrigation and Hydraulics Eng. Mansoura university , Egypt.

Leonards, G.A., (1968), *Les Fondations* (traduction de "Foundation Engineering". McGraw- Hill New York 1962 par un groupe d'ingénieurs des Laboratoires des Ponts et Chaussées). Dunod, Paris, 1106 p

Loudiere, D. (1972). Elément théorique sur le drainage dans les barrages en terre homogènes. C.T.G.R.E.F., Nov, Mallet, Ch. & Pacouant, J. ; (1951). *Les barrages en terre.* Edition Eyrolles.

Oosterbaan, R.J. & Nijland, H.J. (1994). Determining the saturated hydraulic conductivity. *Drainage Principles and Applications.* International Institute for Land Reclamation and Improvement (ILRI), Chapter 12 in: H.P.Ritzema (Ed.) Publication 16, second revised edition, Wageningen, The Netherlands. ISBN 90 70754 3 39

Ritzema, H.P. ; (1994). Subsurface Flow to Drains. *Drainage Principles and Applications.* International Institute for Land Reclamation and Improvement (ILRI), Chapter 8 in: Publication 16, second revised edition, Wageningen, The Netherlands. ISBN 90 70754 3 39

Thieu N.T. M., Fredlund D. G., Fredlund M. D. & Hung V. Q. (2001). *Seepage Modeling in a saturated/unsaturated Soil System.* International Conference on Management of the Land and Water Resources, MLWR, October 20-22, Hanoi, Vietnam.

Volkov, V. (1986). *Ouvrages hydrauliques.* Guide de Thèse, ENSH, Blida, Algeria, pp 120-128. Xu Y.-Q., Unami K. & Kawachi T. (2003). Optimal hydraulic design of earth dam cross section using saturated–unsaturated seepage flow model, Advances in Water Resources, 26: pp 1–7

Zhang J., Xu Q. & Chen Z.,(2001). Seepage Analysis Based on the Unified Unsaturated Soil Theory, Mechanics Research Communications, , 28, No 1:pp 107-112,

06 AN INVESTIGATION ON THE SUBSURFACE DRAINAGE AND WATER QUALITY IN THE JEDDAH-MAKKAH AQUIFER ZONE, LOCATED IN THE WEST CENTRAL REGION OF THE ARABIAN SHIELD IN SAUDI ARABIA

1. INTRODUCTION: AN OVERVIEW

1.1 Ground water in Saudi Arabia

In locations with arid climates, where the average annual rainfall is less than 200 mm, the process of replenishing local and regional aquifers is primarily indirect, extremely restricted, and of little importance (Lloyd 1999). With the exception of the small amount of groundwater found in shallow alluvial aquifers, the majority of stored groundwater in local and regional sedimentary aquifers is non-renewable fossil water. This water has been stored for a long time, with ages ranging from around 10,000 to 32,000 years. Groundwater, both renewable and non-renewable, has been extensively utilized in several nations like the USA, Australia, Spain, India, Jordan, Oman, Libya, Bahrain, UAE, Egypt, and Saudi Arabia to support agricultural and household activities. Saudi Arabia's extensive exploitation of groundwater, including non-renewable fossil water, particularly following the rise in its oil income in 1974, serves as a prime example of the intense utilization of groundwater for irrigated agriculture to support the socioeconomic progress of local people.

The majority of groundwater in Saudi Arabia is located in the abundant, easily penetrable aquifers within the sedimentary basins to the north and east. Additionally, groundwater can be found in the fractured, ancient crystalline rocks of the Arabian Shield. These rocks play a crucial role in facilitating surface runoff and localized, shallow wadi underflow. The northern principal aquifers are composed of various permeable formations from the Early Paleozoic period, which are interspersed with impermeable layers of clay. On the other hand, the eastern aquifers consist of both karstified carbonates from the Tertiary period and clastic formations from the Mesozoic to basal Paleozoic periods. Located south of the Arabian Shelf, there is a single, thick, base Early Palaeozoic sandstone structure that serves as a highly productive aquifer. Groundwater is stored in a complex system of about 20 stratified major and subsidiary aquifers that vary in geological age (Fig. 1) (MAW 1984).

The Arabian Shelf encompasses the extensive sedimentary aquifers, primarily composed of limestone and sandstone, which underlie the underlying basement rocks. It spans approximately two thirds of Saudi Arabia, covering an area of 1,485,000 km2 (MAW 1984). The overall thickness ranges from a few hundred to over 5,000 meters (MAW 1984). The main aquifers include the Saq, Wajid, Tabuk, Minjur, Dhruma, Biyadh, Wasia, Dammam, Umm Er Radhuma, and Neogene. The secondary aquifers include Al-Jauf, Al-Khuf, Al-Jilh, the Upper Jurassic, Sakaka, and the Lower.

The figure 1 shows Cretaceous Aruma basalts and wadi sediments. The quality of groundwater differs among aquifers and across different locations. The isotopic tests indicated that the fossil groundwater present in the aforementioned aquifers has an age ranging from 10,000 to 32,000 years. The sedimentary aquifers contain significant amounts of groundwater (KFUPM/RI 1988). According to hydrogeological studies by KFUPM/RI (1988) and Alawi & Abdulrazzak (1994), the estimated groundwater reserves up to a depth of 300 meters below the ground surface is approximately 2,185 km3. These studies also indicate that the total annual recharge is around 2,762 Mm3.

In dry places, the inconsistency of replenishment results in notable disparities in the quality of the groundwater across different locations and periods of time. The quality of groundwater is heavily influenced by factors such as the chemical makeup of the water-bearing deposit and the amount of time it spends in the aquifer.

In order to effectively plan water resources management, it is crucial to accurately determine water quality based on its spatial distribution and genetic characteristics. This will allow for a comprehensive understanding of the qualitative impacts of both natural and human influences on water quality.

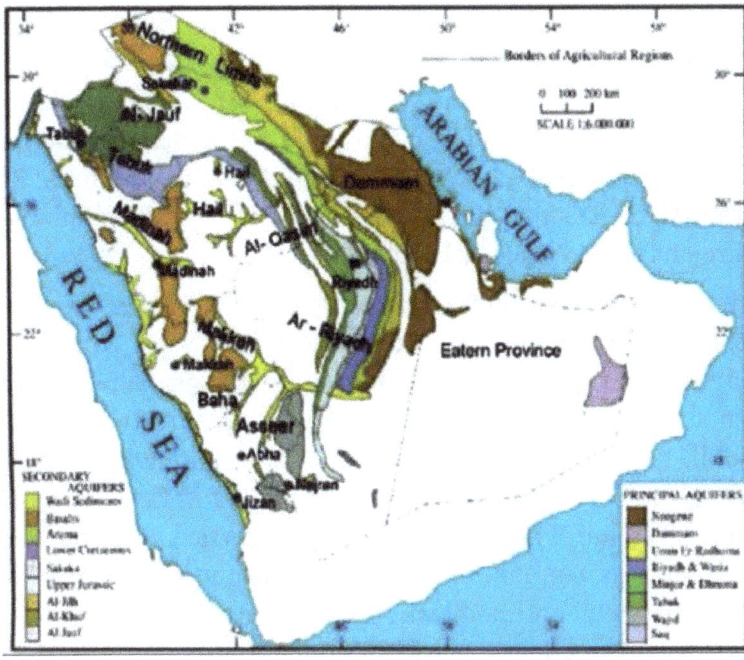

Fig. 1. The extension of the outcrop area of principle and secondary aquifers in agriculture regions in Saudi Arabia.

Therefore, any decline in the quality of groundwater could lead to a significant decrease in the availability of usable resources or provide significant challenges for domestic, agricultural, and other activities. Consequently, there is an increasing demand to analyze the chemical composition of groundwater, namely the levels of main ions and trace elements, and assess how these factors impact its appropriateness for various applications.

1.2 Water supply sources in Saudi Arabia

Over time, the reliance on non-renewable groundwater resources has grown since home and industrial water usage increasingly depends on renewable groundwater and desalination methods. Desalination plants and renewable groundwater are the primary sources for home and industrial water use, with non-renewable groundwater serving as a secondary supplier to meet these demands. This study focused on the hydrogeology and hydrochemistry of the groundwater aquifers in the Jeddah-Makkah Al Mukarramah District, located in West Central Saudi Arabia. The study primarily relies on field observations and measurements, supplemented by several laboratory approaches.

1.3 General climates and geology of the study area

The study area lies within the western province of Saudi Arabia between latitudes 20o 54 00\ and 21o 57 23\ N and longitudes 39o 17 09\ and 40o 00 57\ E (Fig. 2).

The present study will involve the three major drainage basins and their main tributaries. These are as follow: 1) Usfan, including, Haddat Ash Sham, Al Baydah-Mudsus, Ash Shamiyah and As Suqah sub-basins, 2) An Numan, including, Dayqah, Rahjan, Arar, Yarij and Uranah sub- basins, and 3) Fatimah, including, Al Sail Al Kabir, Al Yammaniyyah, Al Shamiyyah, Hawarah, Alaf, Bani Omair-Al Rayan and Al Jumum-Bahrah. All the basins lying partially within the Arabian Shield, while their lower parts are on the Red Sea coastal plain.

Several hydrogeological research activities were carried out in the study area since the early seventies. Most of these works were concerned with the groundwater condition, aquifer characteristics and the groundwater quality in the wadi sediments (e.g. Italconsult, 1976; Al- Khatib, 1977; Al-Hajeri, 1977; Jamman, 1978; Al-Nujaidi, 1978; Al-Gamal and Sen, 1983; Mansour, 1984; Sharaf et al. 1988; Al Kabir, 1985; Basmci and Al-Kabir, 1988; Alyamani and Hussein, 1995; Alyamani et al. 1996; and Alyamani, 1999). A few studies concerned with the trace elements concentrations such as Mn, Pb, Si, Al, F and B in the groundwater are available .i.e. Bazuhair et al. (1992) outlined comprehensive investigations on groundwater condition included water chemistry within Khulais basin. Detailed hydrogeological and hydrochemical studies were carried out in Wadi Usfan by Sharaf et al. (2002). Several detailed studies regarding geological and geomorphological characteristics of the area around An Numan basin are available (e.g. Brown et al. 1963; Zaidi, 1983 and 1984; Moore and Al-Rehaili, 1989).

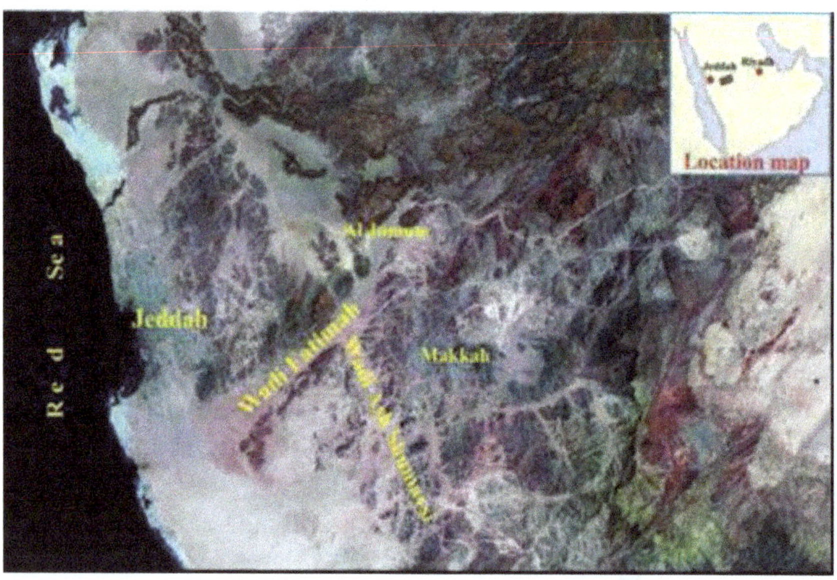

Fig. 2. Satellite image of the study area.

The climatic conditions all over the study area may play an important role in defining the hydraulic response of the watersheds and groundwater quality existing in the region. Generally, the climate is typically arid and the rainfall is irregular and has torrential nature. The rainfall occurs during winter season, while in the autumn and spring the area is subjected to isolated events. The average annual rainfall is about 60 mm in the lowland areas. Where, moving towards the eastern direction the rainfall increases to more than 170 mm/year. Such variation in the rainfall amount can be attributed to the orographic effects of the Red Sea escarpment. Rainfall distribution over the study area was characterized in time and space. Data from the five rainfall gauges were used to establish the rainfall distribution. The variation of rainfall in the spatial dimension reflects the topographic effects, since the highest values were recorded at the highest stations. The average monthly rainfall was computed in order to give an approximate idea of the seasonal variation in the local rainfall patterns. All effective rainfall is concentrated between November and April (with very minor exceptions in early May).

From the geologic points of views, the study area comprises Precambrian-Cambrian basement complex, Cretaceous-Tertiary sedimentary succession, the Tertiary- Quaternary basaltic lava flows, and the Quaternary-Recent alluvial deposits (Fig. 3). The Precambrian rock units in the study area consist of Late-Proterozoic basaltic to rhyolitic volcanic and volcanoclastic and epiclastics of primitive island-arc type, that have been multiply deformed and metamorphosed and injected by intrusive bodies of different ages and compositions. These rock units are divided into Zibarah, Samran, and Fatimahh groups (sedimentary rocks). Plutonic rock units are gabbro, diorite, tonalite and granodiorite to monzonite of probably early Cambrian ages.

The Cretaceous-Tertiary sedimentary succession is exposed beneath a cover of flat-lying lavas and Quaternary deposits in the study area. This succession is subdivided into the Haddat Ash Sham, Shumaysi, Khulays, and Buraykah formations. It consists of clastic rocks dominated by sandstones, shale, mudstones, oolitic ironstones, and occasionally conglomerates. A middle Cretaceous age has been assigned to Haddat As Sham Formation. Basalt lava flows form discontinuous caps overlying the upper levels of both the basement complex and the sedimentary rocks; the lavas are either rest on peneplain or infilled ancient wadis. They are preserved in three north- northwest trending, asymmetric depositional troughs which are the Sham, Suqah and Shumaysi troughs. These troughs are bounded in the north by faults downthrown to the west and in the west by an unconformity at the base of the easterly dipping strata.

The research area is extensively covered with Quaternary sediments. They primarily exist in the major catchment areas of Haddat Ash Sham. The primary components of the Quaternary rocks consist of terrace gravel, alluvial fan deposits, talus deposits, alluvial sands and gravels found in wadi beds, as well as certain eolian structures. The thickness of these deposits exhibits significant variation across different locations.

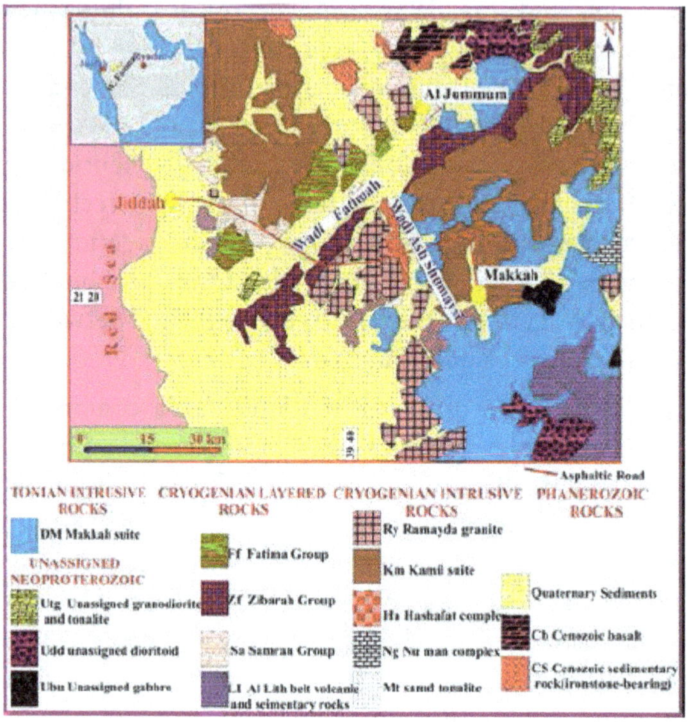

Fig. 3. Geologic map of Makkah district including Wadi Usfan, Wadi Fatimah and wadi An Numan (Petter Johnson, 2006).

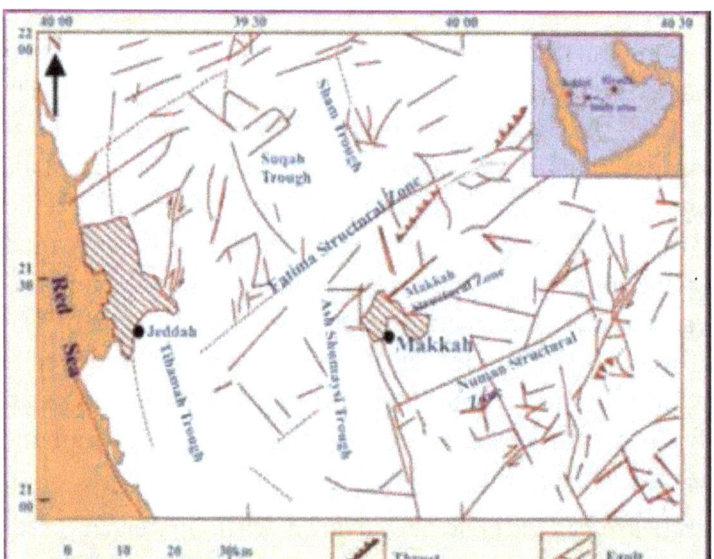

Fig. 4. General structural geologic map of Makkah district including W. As Suqah area (after Sharaf, 2011b).

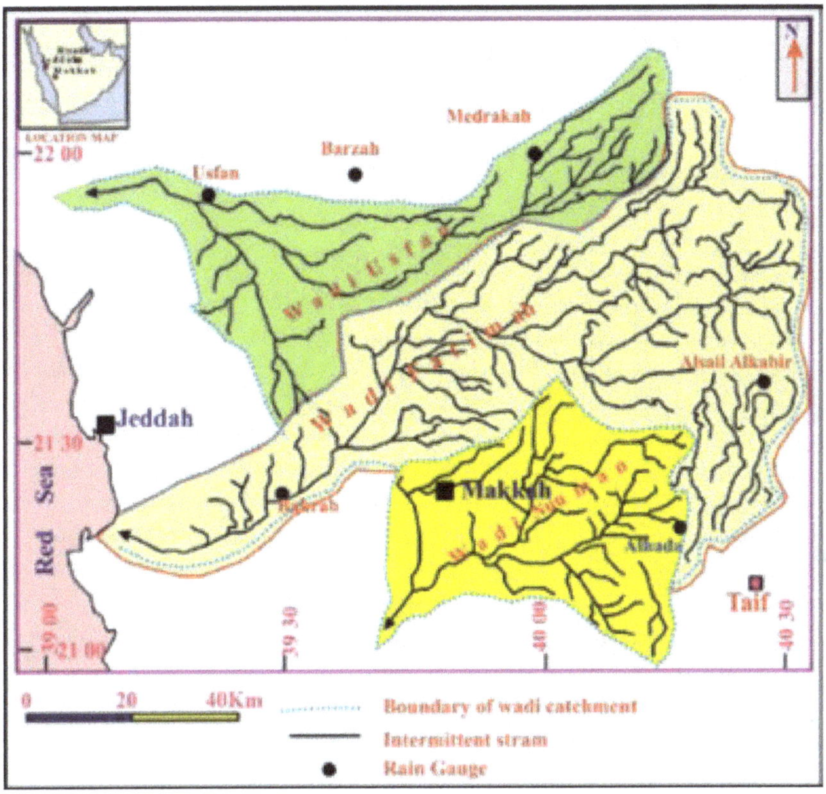

Fig. 5. Drainage basin map of the studied basins (Sharaf, 2011a).

Three main sets of faults: NW, NE and N (Fig. 4). The NW faults are the oldest and seem to have controlled the depositional troughs in the study area, mainly those of Haddat Ash Sham and As Suqah. They are mostly normal faults dipping steeply to the southwest. The NE faults displace the NW set and seem to be second component in block faulting. The N trending faults are shear faults with lateral displacement for the above mentioned NW and NE sets.

The research area consists of three separate geomorphologic zones: the hilly region in the far east, the pediment region, and the coastal plain (Fig. 5). The hilly region consists primarily of folded and faulted Precambrian rocks known as the Hijaz Highlands, as well as Tertiary lava flows referred to as Harrat Rahat. They constitute a continuous vertical segment that spans from the northern to the southern region. The pediments define the mountainous area to the west and are characterized by their elongated characteristics, which extend in a westward direction and occasionally resemble little protrusions.

The pediments are characterized by a thin layer of alluvial and aeolian sediments that overlay a heavily worn surface of Precambrian rocks. The hilly region primarily serves as a recharge area, whereas the pediment region might be classified as an area of surface flow.

Table 1. Average monthly rainfall for different rain gauges (mm) (Sharaf, 2011a).

STATION NAME	Jan	Feb	Mar	Apr	May	Jun	Jul	Aug	Sep	Oct	Nov	Dec	Elevation m.a.s.l
Alhada (1980-1992)	22.4	7.0	18.3	19.6	30.4	2.5	0	8.0	7.6	14.4	21.2	36.7	1940
AlsailAlkabir (1982-1993)	5.3	0.4	5.7	17.7	12.0	1.6	0.2	0.9	2.6	1.7	7.6	10.8	1230
Bahrah (1966-1997)	18.0	4.0	1.5	4.2	1.2	0	0	2.0	1.5	0.9	11.4	18.0	116
Barzah (1976-1997)	8.5	3.9	7.0	7.0	0.8	1.9	0.7	2.7	5.5	1.1	18.0	15.0	350
Medrakah (1966-1996)	16.6	3.9	7.7	23.2	9.0	1.1	4.0	3.5	13.6	7.9	12.9	7.1	710

1.4 Data of recharge area

The rainfall data collected from five rain gauges stations; these are Alhada, Alsail Alkabir, Bahrah, Barzah, and Medrakah. The collected data are in the form of daily rainfall for different number of years. These data were collected from the Hydrology section of the Ministry of Agriculture and Water (MAW) and from Meteorology and Environmental Protection Administration (MEPA). A total of 400 groundwater samples were collected from the privately owned drilled wells within the studied basins. Most of the groundwater samples were taken from intensively pumped wells in order to avoid any local contamination or change in chemistry caused by evaporation or gas exchange in the well itself rainfall distribution over the study area was characterized in time and space. Data from the five rainfall gauges (Fig. 5 & Tab. 1) were used to establish the rainfall distribution. This table shows that the variation of rainfall in the spatial dimension reflects the topographic effects, since the highest values were recorded at the highest stations (Medrakah, Alsail Alkabir, and Alhada, Fig. 5). The average monthly rainfall was computed in order to give an approximate idea of the seasonal variation in the local rainfall patterns. All effective rainfall is concentrated between November and April (with very minor exceptions in early May).

2. WADI USFAN

From the geologic point of view, the study area comprises Precambrian-Cambrian basement rocks, Cretaceous-Tertiary sedimentary succession, the Tertiary-Quaternary basaltic lava flows, and the Quaternary-Recent alluvial deposits. Wadi As Suqah is a NW-SE low lands surrounded from the west by Precambrian rocks overlain by black basaltic lava flows (Harrat, Fig. 6, 7).

Precambrian rocks of wadi Usfan have been classified by Moore and Al-Reheili (1989) into Late- Proterozoic basaltic to rhyolitic volcanic and volcanoclastic and epiclastics of primitive island-arc type, that have been deformed and metamorphosed during many times and injected by intrusive bodies of different ages and compositions.

Sedimentary rock units of Usfan basin have been classified by Petter Johnson (2006) into: 1- The Cryogenian layered rocks which are represented by the Samran Group (sa), 2- The Cryogenian intrusive rocks which are represented by The Hishash granite (ig) which intrudes the Kamil Suite and Samran Group and composed of monzogranite and subordinate granodiorite (Moore and Al-Rehaili, 1989) and the Kamil Suite (km) which consists of mafic, intermediate, and felsic plutonic rocks of calc-alkalic and locally trondhjemitic affinities and, 3- The Edicarian layered rocks are represented by Shayma Nasir Group (Sn), which includes polymict conglomerate; basaltic, andesitic, dacitic, and rhyolitic lava, tuff, and agglomerate; and red-brown arkosic, volcaniclastic, and calcareous sandstone.

Cenozoic rock units of Usfan Basin (CS) are exposed beneath a cover of flat-lying lavas and Quaternary deposits. Brown and others (1963) introduced the names Shumaysi and Usfan formations to these sedimentary sequences after Karpoff (1958). Spincer and Vincent (1984) divided the Shumaysi Formation into the Haddat Ash Sham, Shumaysi, Khulays, and Buraykah formations.

Sharaf (2009 a and b) carried out a geophysical and hydrochemical study on Haddat Ash Sham and Ash Shamiyah areas and concluded that, the groundwater is mainly present in the Quaternary alluvial deposits and the Tertiary sandstones and conglomerates of Haddat Ash Sham and Ash Shumaysi Formations. In As Suqah area (Fig. 6), the groundwater is present within two main water-bearing units: the alluvium of the Wadi system under unconfined conditions and within the clastic layers of the Cretaceous sedimentary succession. The general groundwater flow in the aquifer system follows the surface drainage towards the northwest to Usfan city; it takes place from areas of high potential to areas of low potential (Fig. 8). Both the elevation and pressure heads define the direction of groundwater flow towards northwest. The values of the depth to water in Table 1 represent the absolute depth to the water while the values of the contour lines showing the flow direction in Fig. 7 represents the water depth above sea level (a. s. l.).

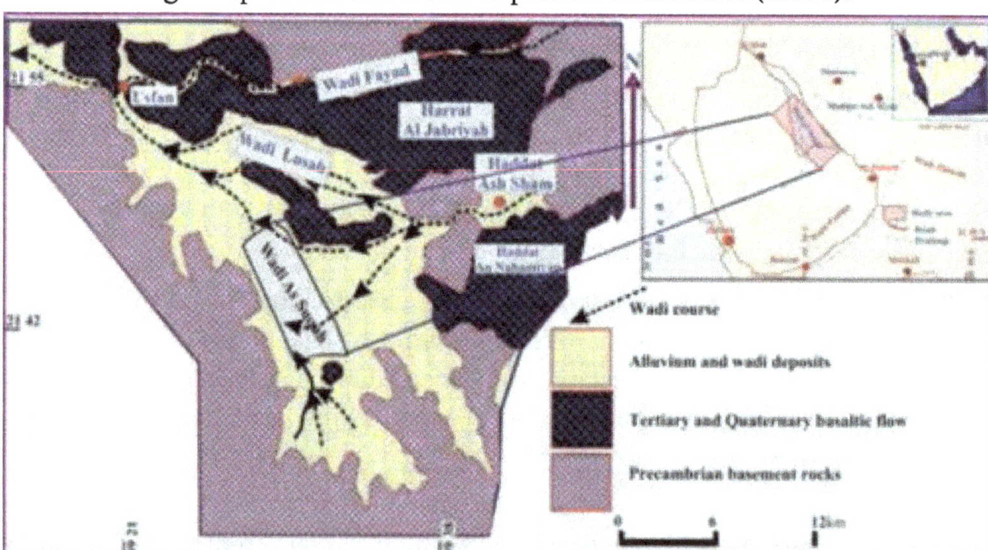

Fig. 6. Simplified geologic map of Usfan basin (Sharaf, 2011b).

In Usfan basin, the geological and geophysical exploration for groundwater is based mainly on integrated electro-resistivity (VES, Fig. 9), and four drilled test wells (Fig. 10) and seismic and magnetic geophysical tools. The results of theses exploration tools revealed the following conclusions: 1) Groundwater occurs mainly in two water-bearing horizons, the alluvial deposits and within the clastic sedimentary rocks of Haddat Ash Sham and Ash Shumaysi formations. The shallow zone is characterized with a saturated thickness of 3- 20 m and water is found under confined to semi-confined conditions, 2) Water levels were encountered at depths varying from 3-16m in the alluvial wadi deposits, and from 18-62 m in the sedimentary succession, 3) Groundwater movement is towards the west and northwest, following the general surface drainage system, and 4) Hydraulic gradient varies greatly from one point to another depending on the pumping rates and cross-sectional area of the aquifer in addition to its transmissivity.

The locations of the water wells of wadi Usfan (Fig. 7A) show the localization of these wells in the southeastern (upstream) part of the wadi. The cation composition varies between almost exclusively Na+ to dominantly Ca^{2+} and Mg^{2+} with relatively lesser amount of K+. Among the anions, Cl- is dominant. They also reflect that HCO^{3-} and K^{+} ions seem to be rather uniform distribution. The ionic concentration of the groundwater samples has the following general pattern:

$$Na^{+} > Ca^{2+} > Mg^{2+} > K^{+} \text{ and } Cl^{-} > SO_4^{2-} > HCO_3^{-}$$

On the other hand, the spatial distributions of the major ions shown that the major constituents followed the general trend of the groundwater salinity (EC) in the basin.

The chemical analyses of the trace elements shown that, each element has its own dispersion in the area without any general trend of distribution. Such pattern occurred reflecting that most of these elements are probably locally affected as a result of chemical reactions between groundwater and the parent rocks. However, the majority of these elements were marked in As Suqah and Ash Shamiyah regions, such as B, As, Al, Mn, Li, Cd, Ba, Cr, and Mo. The high concentrations of these elements is almost present adjacent to the basaltic lava in the middle part of the study area, reflecting that the chemical weathering of basalt may consider as a source of them. Haddat Ash Sham, on the other hand, characterized by high contents of Hg, Cu, Co, F and PO4.

In Wadi Usfan, the water quality is highly variable. The groundwater salinity is non- uniform aerially, and there are wide differences among the individual wells as well as the sub-areas. Mean values and ranges of the EC measurements for each sub-area (i.e. Haddat Ash Sham, Ash Shamiyah, Al Baydah-Mudsus and As Suqah) are shown in Table 2. Both Haddat Ash Sham and Ash Shamiyah sub-basins are characterized by relatively low saline water, whereas, highly mineralized water almost marked Wadi As Suqah sub-basin where the groundwater salinity reached up to 30,000 μS/cm with an average of about 18800 μS/cm. The EC measurements of the groundwater within Haddat Ash Sham and Ash Shamiyah sub-basins indicated that few wells yield relatively low saline water (< 1500 μS/cm). These are distributed randomly and are often near wells providing moderately mineralized water. It also observed that within these two areas, the groundwater from wells drilled in the center of the wadi alluvium are not as mineralized as the waters extracted from wells located along the edges of the wadi channel. The EC and SO4 (Fig. 7B, C respectively) distribution shows that the groundwater salinity almost increased towards the downstream part of the wadi, which probably matches the general flow of the groundwater.

Table 2. Average groundwater salinity in Usfan sub-basins.

Basin Name	Sub-Basin Name	Maximum	Minimum	Mean
WADI USFAN	Haddat Ash Sham	8600	875	3699
	Ash Shamiyyah	10890	1808	6482
	Mudsus-Al Baydah	11940	2200	5504
	As Suqah	30800	7720	18837

The dominant water types are NaCl and CaCl2. On the other hand, a few wells yield medium to high saline water. Those are concentrated in Wadi Haddat Ash Sham, Ash Shamiyah and Al Baydah-Mudsus sub-basins.

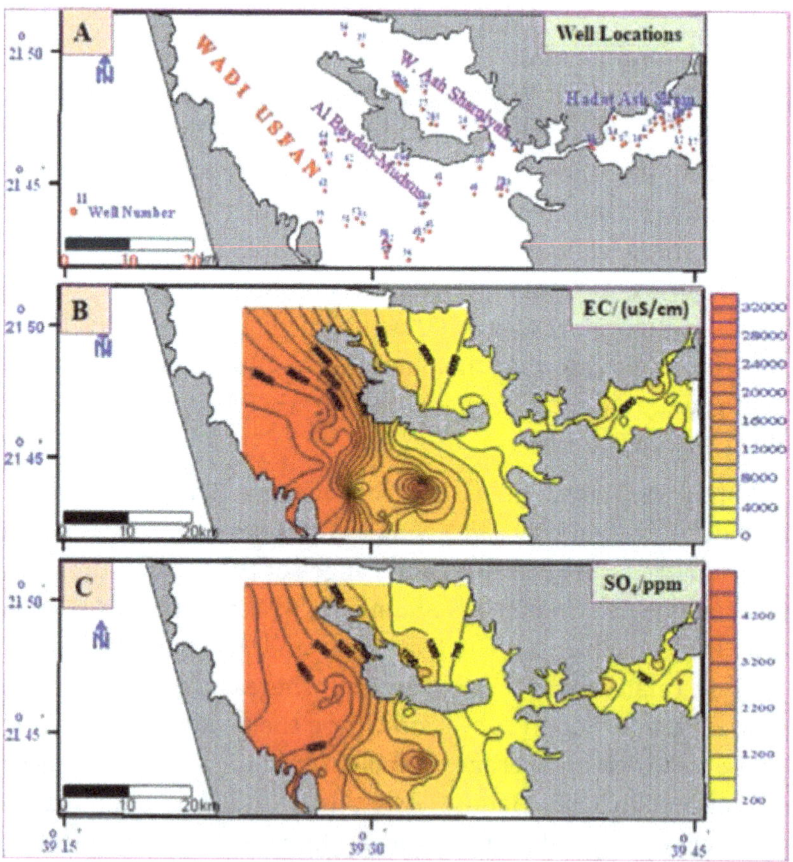

Fig. 7. A= Well locations in wadi Usfan; B= EC; and C= SO4 content of the groundwater of wadi Usfan.

These are fallen within the satisfactory limits and can be used safely for agricultural practices. In contrast, Wadi As Suqah sub-basin shown highly saline water, although they reflected low to high SAR values. Under this condition, the groundwater tends to restrict its utilization for agricultural activities. The concentration of NO^{3-} ions is relatively high. The major sources of NO^{3-} probably as a result of intensive use of chemical fertilizers in these regions. Poultries may also considered as another source of NO^{3-} particularly in the upper part of Wadi Ash Shamiyah in Usfan basin.

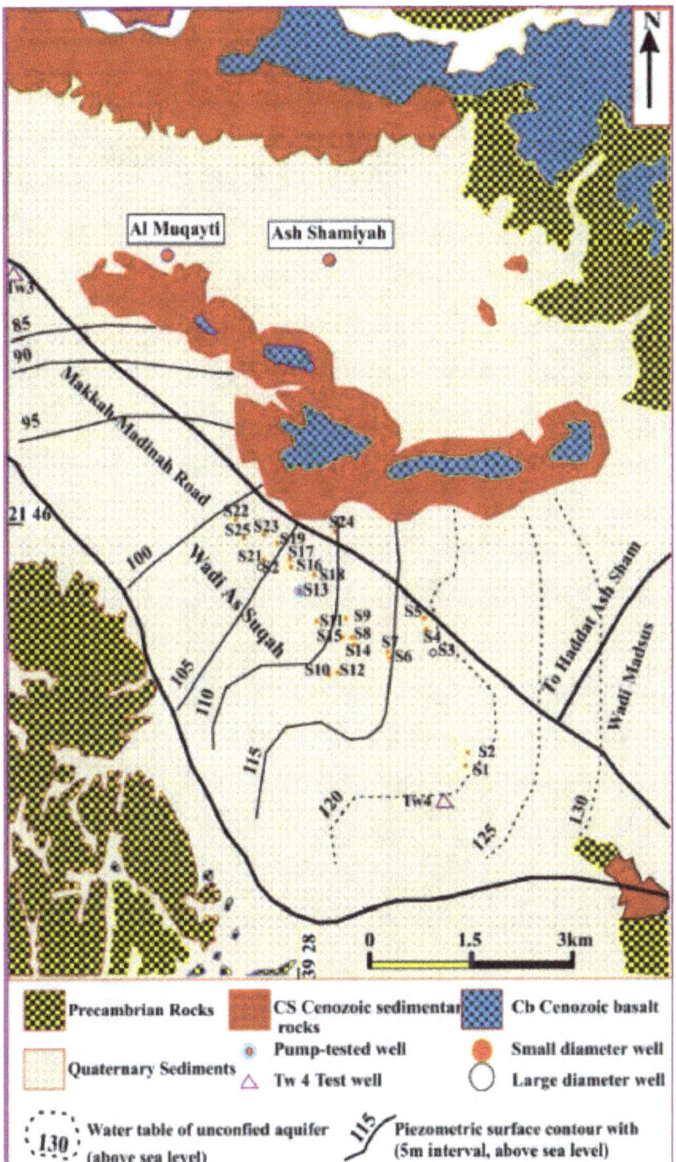

Fig. 8. Detailed geologic map showing the locations of the studied boreholes of As Suqah area (Sharaf, 2011d).

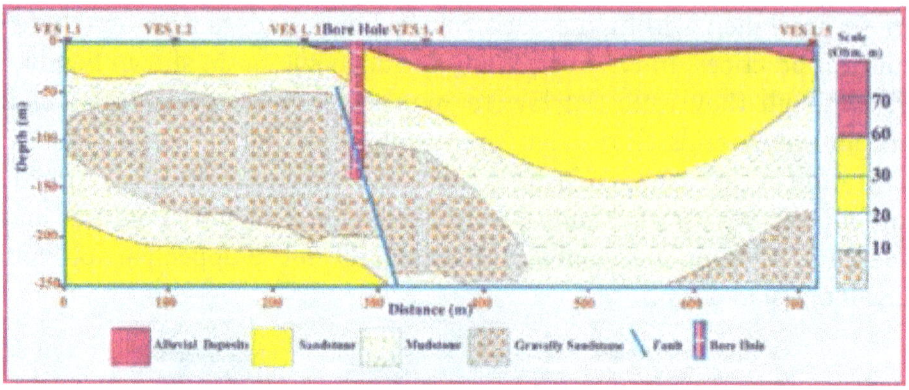

Fig. 9. Haddat Ash Sham (Usfan Basin) geo-electrical cross-section (Sharaf, 2011b).

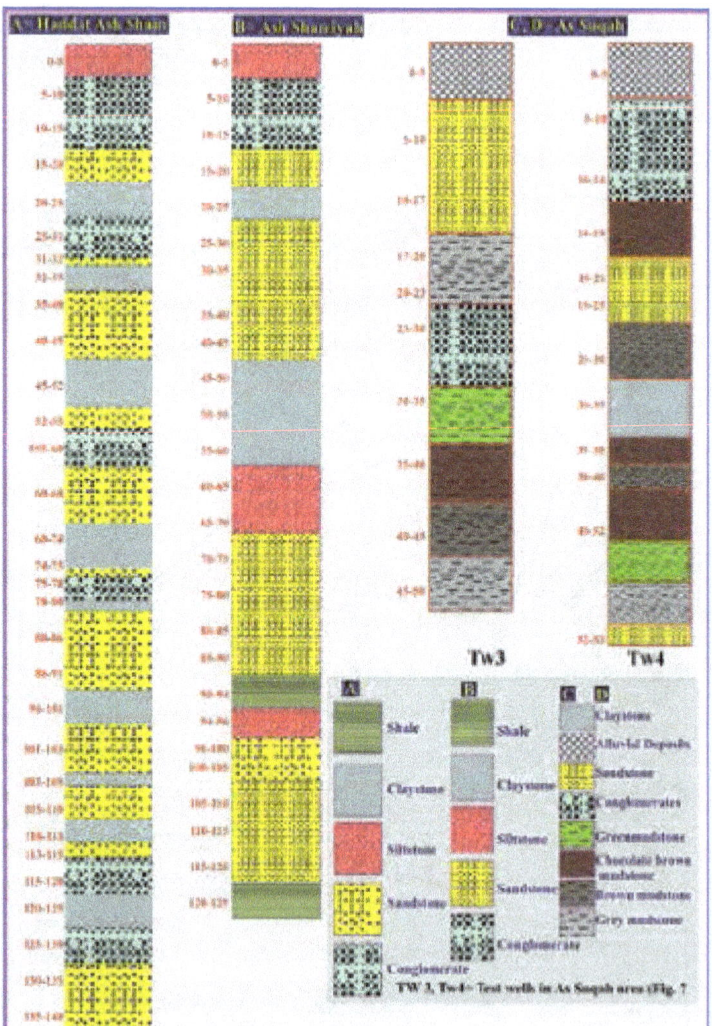

Fig. 10. Stratigraphic logs of the drilled test wells in Usfan basin (Modified from Sharaf, 2010a, b and 2011b).

3. WADI AN NUMAN

An Numan basin primarily originates in the Hijaz Highlands (Fig. 11). The drainage of the basins is typically well-developed and follows a dendritic pattern, as shown in Figure 5. The research area consists of three separate geomorphologic zones: the mountainous region in the extreme east, the pediment region, and the coastal plain. The hilly region consists primarily of folded and faulted Precambrian rocks known as the Hijaz Highlands, as well as Tertiary lava flows called Harrat Rahat. They constitute a contiguous vertical segment that spans from the northern to the southern direction. The pediments define the mountainous area to the west and are characterized by their elongated characteristics, which extend in a westerly direction and occasionally resemble little knobs. The pediments are characterized by a thin layer of alluvial and aeolian sediments that overlay a heavily worn surface of Precambrian rocks. The mountainous region primarily serves as a recharge area, whereas the pediment region might be regarded as an area of surface flow.

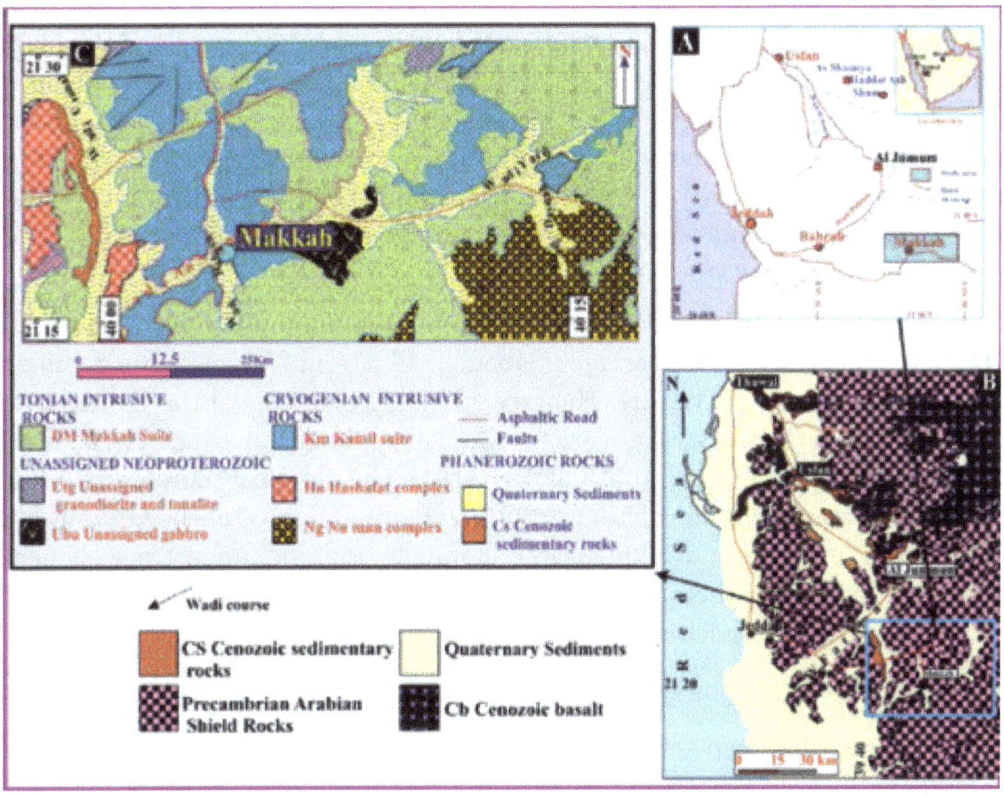

Fig. 11. Geologic map of the study area (Modified after Johnson, 2006 by Sharaf, 2011a).

3.1 Water hydrochemistry

The groundwater aquifer in the wadi An Numan basin and its sub-basins is found in both the Quaternary wadi fill deposits and the weathered fractured crystalline

Precambrian bed rocks. The Quaternary alluvial deposits comprise conglomerate consisting of well-rounded pebbles to boulders. These deposits are primarily documented in wadis with steep sides. In these regions, the sedimentary deposits are shallow, whereas in the lower sections, they are deep and consist of gravel that is reasonably well arranged and sand that is of medium to fine texture. The water depth maps of wadi An Numan indicate that the wells located in the uppermost tributaries, which are the regions where water is replenished, are relatively shallow. On the other hand, the wells situated within the main channel of wadi An Numan are deeper. The minimum water depth in the downstream of the main channel of wadi An Numan is 25m, while the maximum water depth is approximately 39m.

3.1.1 Sampling and analytical methods

A total of 65 groundwater samples were collected from the privately owned drilled wells within Wadi An Numan basin. During the field survey and sample collection, groundwater temperature, electrical conductivity (EC) and pH were measured at the well sites. Most of the groundwater samples were taken from intensively pumped wells in order to avoid any local contamination or change in chemistry caused by evaporation or gas exchange in the well itself. These water samples were analyzed by using Inductively Coupled Plasma-Mass Spectrometry (ICP-MS) in the laboratories of the Faculty of Earth Sciences, King Abdulaziz University, Jeddah, Saudi Arabia. The chemical analyses were performed for the cations Na^+, K^+, Ca^{2+}, Mg^{2+} and the anions HCO_{3-} , CO_3^{2-}, SO_4^{2-}, Cl- , NO_{3-} .

In Wadi An Numan, the groundwater salinity is rather low with EC ranged from 542 to 5400 µS/cm with an average of about 1539 µS/cm (Table 3). The dominant water types are NaCl and CaCl2. The rare and trace elements of the groundwater samples were determined. These are As, Mo, Zn, P, Pb, Co, Cd, Ni, Ba, Fe, B, Si, Hg, Mn, Cr, V, Bi, Cu, U, Al, Li, Rb, Au, Cr, Sr, F and NH4. The chemical analyses show that, the groundwater is enriched with nitrate (NO_3) ions and most of the groundwater is very hard for use as a drinking water supply. In addition, a few trace elements showed an increase in their contents and exceed the maximum acceptable limits of the standards that restricted to its utilization for drinking water.

3.1.2 Description of the groundwater chemistry

3.1.2.1 Durov diagram

The main purpose of the Durov diagram is to show clustering of data points to indicate samples that have similar compositions (Hem, 1989). The durov diagram of the chemical analyses of the water samples of the different areas (Fig. 12) revealed that, the water types of wadi Dayqah sub-basin are mainly of CaCl2 types except some analyses which are of calcium bicarbonate types. Analyses no. 5, 7, 9, 10 are of calcium sulphate types. The chemical analyses of the main channel of wadi An Numan sub-basin are of calcium chloride and calcium sulphate types with less frequent calcium bicarbonate types. Some analyses are of NaCl and others are of sodium sulphate water type. In wadi Arar sub-basin the water types are mainly calcium chloride, calcium sulphate and calcium bicarbonate. In wadi Rahjan sub-basin the water types are of calcium chloride and calcium sulphate with less frequent Na and Ca bicarbonates and Ca sulphates. From the Durov diagram we notice that, the analyses of the main channel of wadi An Numan are enriched in Cl and SO_4.

3.1.2.2 Trilinear diagram

The plotting of the analyses of the different sub-basins on the trilinear diagram (Fig. 13) revealed that, the groundwater within the study area is predominantly a mixture of calcium-sodium ions and the upper tributaries of Wadi Numan contains fresh water while downstream (main channel of wadi An Numan, Wadi Rahjan and wadi Arar) waters are brackish. This is due to the very slow interaction between the fractured crystalline rocks and the rainfall waters in the upstream (fresh water) while in the downstream, the water is interacted with the alluvial sediments as well as the highly weathered and fractured bed rocks which lead to the addition of CO3, Na, Ca, K and Mg to the water with time leading to the formation of saline and brakish waters of HCO_{3-} -Ca_2^+ - Na^+ -SO_4^{2-} -Cl -Ca_2+ type.

Table 3. Statistics of the chemical analyses (65 samples, concentrations in mg/l) (Sharaf, 2011a).

Item	Minimum	Maximum	Average	St. Dev.
Ca	66	432	154.766	72.155
Mg	3.26	160	36.893	27.756
Na	37.2	500	101.466	77.288
K	2.57	14.5	6.753	1.735
HCO_3	127.9	392.5	214.36	54.221
SO_4	89.0	1350	262.252	205.664
SO_3	74.2	1125	219.007	171.415
NO_3	6	560	78.32	83.5
Cl	44	1085.8	205.075	173.17
Sum Cations	5.943	55.894	15.344	8.814
Sum Anions ○_	5.579	55.945	14.758	9.099
Alkalinity	11.74	36.028	19.676	4.977

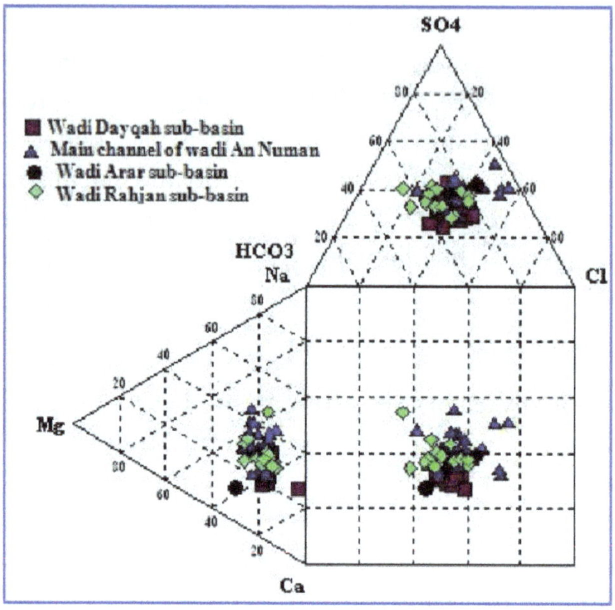

Fig. 12. Durov Diagram of the chemical analyses of the water samples of the different sub- basins of wadi An Numan (Sharaf, 2011a).

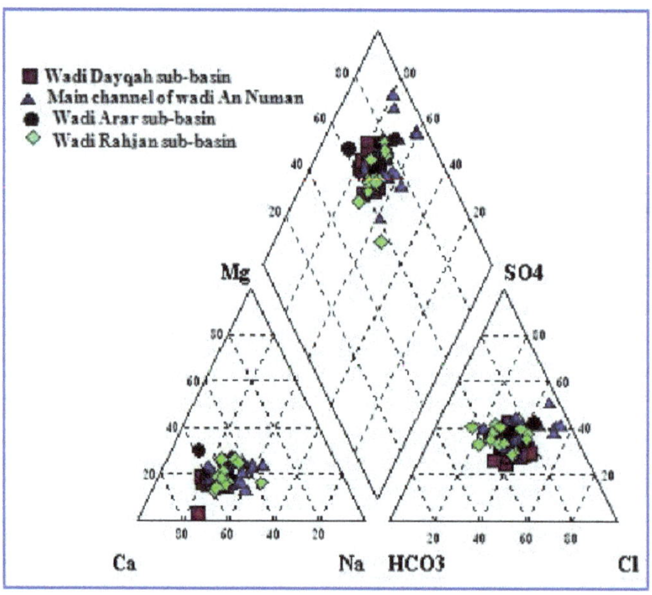

Fig. 13. Trilinear diagram of the chemical analyses of the ground water samples of the different sub-basins of wadi An Numan (Sharaf, 2011a).

3.2 Suitability of groundwater

3.2.1 Irrigation water

There are many factors that determine whether groundwater is suitable for irrigation use, which include: 1) the total salt concentration of the water, 2) the concentration of certain ions that may be toxic to plants or that have an unfavorable effect on crop quality; and 3) the concentration of cations such as Na+, Ca2+ and Mg2+ that can cause deflocculation of the clay in the soil and resulting damage to soil structure and declines in infiltration rate. The quality requirements of irrigation water or universal standards for it cannot be formulated, and what might be a poor water at one place could be quite acceptable somewhere else.

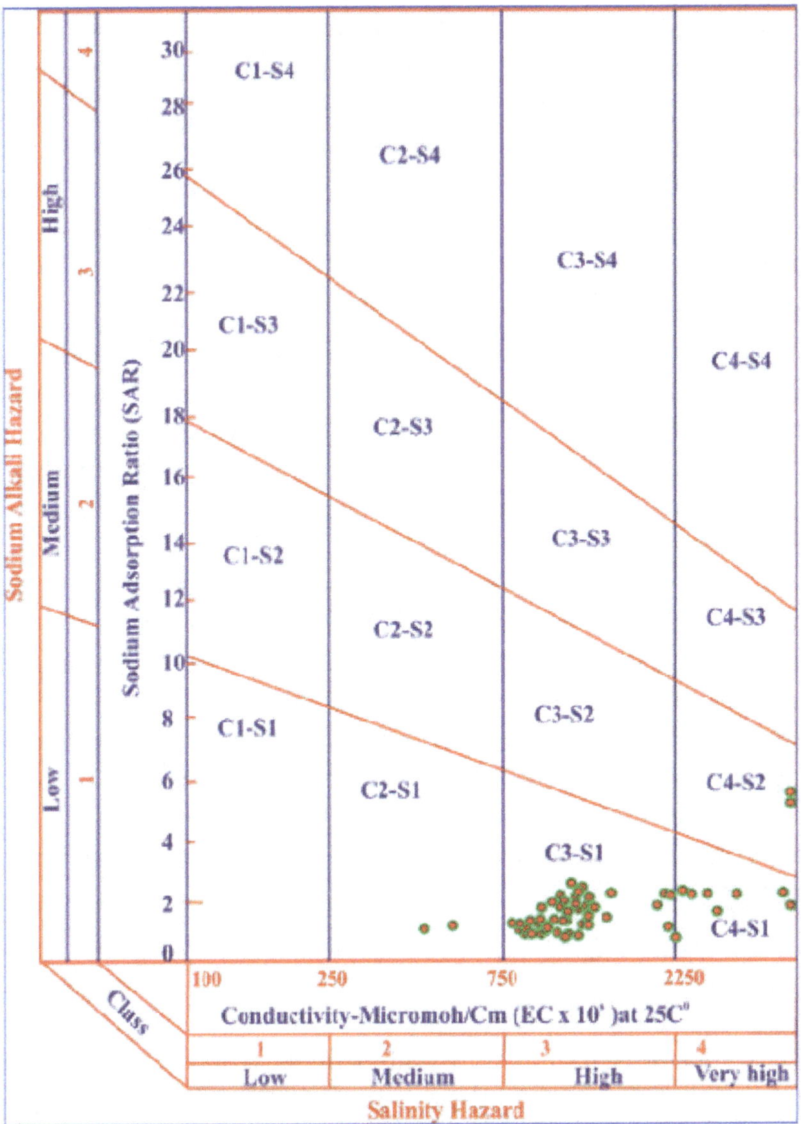

Fig. 14. US Salinity classification of wadi An Numan groundwater for irrigation (USLL, 1954) (Sharaf, 2011a).

Most of the groundwater in Wadi An Numan are fallen within the satisfactory limits. The calculated values of SAR for the groundwater in Wadi An Numan are low according to the recommended water classification for SAR (Lloyd and Heathcote,

1985). The minimum SAR values of wadi An Numan groundwater is 0.8 while the maximum is 5.28 and the average is 1.85. Wadi An Numan groundwater samples are of SAR values between values 1 and 3 and EC values between 1000 and 3000 (Fig. 14). Most of the samples are located within the C3-S1 and C4-S1. Less frequent samples are present in the fields C2-S1 and C4-S2.

The RSC values for the different sub-basins of wadi An Numan are negative (< 1.25). This is because HCO_{3-} is not an important anion compared to Cl- and SO4 in the groundwater. However, the RSC values in the range found can be used safely for irrigation purposes (Lloyd and Heathcote, 1985). The magnesium hazard (MH) values are > 50 which are considered to be harmful. In Wadi An Numan, all the groundwater samples are fall within accepted limits which in turn the water can be used safely. Boron is essential for growth of plants in very small concentrations, and becomes of toxic effect when present above the optimum level. Boron in Wadi An Numan groundwater is fallen with satisfactory limits.

Although evaporation probably has the greatest influence on the total dissolved solids (TDS) in the investigated regions, the actual chemical composition is also influenced by other factors. These include the net addition of the major constituents such as Ca^{2+}, Mg^{2+}, K, HCO_3 and trace elements by weathering of silicate minerals. The Precambrian and intrusive rocks commonly contain the major rock-forming minerals such as hornblende and biotite that enrich in Ca^{2+}, Mg^{2+}, K+ and Fe. Na+ is added by this means but not that much of these ions. Oxidation and dissolution rocks of gabbroic composition will contribute trace elements such as Ni, Co, Cu, Zn, Cr, Li, Cd, V and Mn. The source of these trace elements will be associated with the major cations such as Ca^{2+}, Mg^{2+}, Fe, N^{a+}, Al and Si which are considered the major cations in the rocks of gabbroic composition. Whereas, in the weathering granitic rocks will contribute trace elements of Cu, Ba, Rb, Pb, As, Mo, U and Br. These elements will be associated with the major ions such as Na^{+}, K^{+}, Al and Si. However, the released trace elements almost occur as lattice substitutes impurities or in solid solution.

Evaporation, recycling of irrigation water and chemical weathering reactions of silicate minerals are the dominate processes affecting the groundwater's chemical composition. The first two processes possibly worked collectively and lead to precipitate evaporitic salts in the irrigated fields around the production wells. Calcite, dolomite and gypsum are the dominant evaporitic salts. The groundwater in the study area seems to be suitable when compared with FAO quality criteria for irrigation. The calculated values of SAR, RSC, and Magnesium hazard indicate well to permissible use of ground water for this aim.

4. WADI FATIMAH

Wadi Fatimah is shallow alluvial and fracture bedrock aquifer in western part of Saudi Arabia, and considered as one of the major source of water supplies in the area. Because of the rapid growth of population, long time of aridity and intensive exploitation of groundwater resources from this wadi has led to concern about potential water quality impacts. Alyamani (2007) studied the nitrate concentration in the upper reaches of Wadi Fatimah. His study shows that the nitrate concentration in the groundwater exceeded the maximum contaminant level due to the effects of cesspool system in the upper reach of the basin.

Wadi Fatimah basin comprises the most important drainage system in the western province of Saudi Arabia (Fig. 5). It considered the major important sources of groundwater to the cities of Makkah, Jeddah and the surrounding villages and towns. The Red Sea escarpment (Hijaz Highlands) and Harrat Rahat are considered the recharge areas of these basins. Generally, the main course of the basin follow westerly to southwesterly directions towards the Red Sea coast. The drainage of the basins generally is well developed and the pattern is typically dendritic (Fig. 5). Although the local relief generally less than 400 meters above the wadi floor, the wadi networks have many small and large tributaries. They are deep and narrow and their longitudinal profiles are rather gentle and in some areas become irregular. They almost follow N-S, NW-SE and SE-NW directions, which probably controlled by major structures, and smaller faults that commonly control lesser drainage channels. The upper stream parts of the basin is rather narrow, where the width of the main wadi course vary from less than 100 m to more than 1.5 km further downstream, where the alluvium deposits are widespread and rather thicker.

Groundwater samples of the main stream were taken from the alluvial and fracture aquifer in winter, 2004. Trace elements were analyzed in each sample for a total of 17 water quality descriptors (variables), these elements are: As, Zn, P, Pb, Ba, B, Si, Al, Li, Cu, Fe, Hg, Mn, Cd, V, and Rb. Among these, only pH was measured in the field. Samples were analyzed in the laboratories of the Faculty of Earth Sciences, Saudi Arabia. The accuracy of this analytical technique was controlled using appropriate standards. Statistical summary including the mean, standard deviation, coefficient of variation, and skewness of trace elements are shown in Table 4.

The analysed trace elements of wadi Fatimah (Table 4) show the following distribution: In the study area, according to Sharaf and Subyani (2011) the average Al content is 0.062mg/l which is very low when compared with WHO guideline of Al (0.2 mg/l). The concentration of Al will depend on the availability and extent of weathering of the aluminosilicates such as clays, pyrophyllites, feldspars, micas and other related minerals. High concentration of Al in soils and groundwater may also be attributed to a number of factors including the availability of dissolved organic matter and fluoride (Edmunds et al., 1992). Arsenic (As): which is the most important element affecting the groundwater quality. The average value of arsenic is 0.015 mg/l which similar to that recommended by the WHO. The average barium content of wadi Fatimah area is 0.208mg/l which is similar to the WHO (0.3 mg/l). The maximum value of boron is 8.54 mg/l while its minimum value is 0.372 and the average is 1.6126mg/l which is higher than that recorded in the normal groundwater and also higher than the WHO (1993) recommended values (0.3 mg/l). The toxic effect for B in humans is found to occur above 20 mg /l-1 (Bolt and Bruggenwert, 1978). Boron usually occurs as a nonionized form as H3BO3 in soils at pH less than 8.5, but above this pH, it exists as an anion, B(OH)4 -1. It is very soluble in soils and can be leached especially in sandy soils (Brady, 1974). It is also dispersed in the environment through fertilizer application. Since the pH was less than 8.5 for all the samples, it is more likely that B would be in the non ionized than the ionized state.

The cadmium in wadi Fatimah groundwater is higher than that recommended by the WHO guideline. The average of Cd is 0.004 mg/l. Cadmium is used normally as a coating material, paint pigment, in plastics, fungicide and is a constituent of some fertilizers. Copper, Cd and Zn exist in aqueous form in the predominately +2 state. Generally, Zn exists as Zn2+ state in pH below 7.7 (Liptrot, 1989). Cadmium (II) oxide occurs in some Zn ores which suggests that the two have identical properties such as that both have the + 2 being the most stable oxidation state (Liptrot, 1989).In natural waters, Cu, Cd and Zn occur in the +2 form. The average cupper content of wadi Fatimah is 0.021 mg/l which is very low when compared with that of WHO (2 mg/l). The average Fe content is 0.032 mg/l which is lower than the recommended WHO iron level (< 0.3mg/l). Iron dissolved in groundwater is in the reduced ferrous iron form. This form is soluble and normally does not cause any problems by itself. Ferrous iron is oxidized into ferric iron on contact with oxygen in the air or by the action of iron related bacteria. Ferric iron forms insoluble hydroxides in water. These are rusty red and cause staining and blockage of screens, pumps, pipes, reticulation systems etc. If the iron hydroxide deposits are produced by iron bacteria then they are also sticky and the problems of stain and blockage are many times worse.

Lead (average 0.001 mg/l) and it is relatively higher than the WHO (0.01). Lead on the other hand is used in automobile industries in form of tetramethyllead and tetraethyllead. Manganese (Mn) is present in wadi Fatimah by very low content (average 0.020 mg/l) and it is relatively lower than the WHO (0.01). Mercury (Hg) is important for the water quality. The average Hg in wadi Fatimah area is 0.018mg/l (Table 3) which is relatively higher than the WHO values.

Table 4. Statistical summary for trace elements (mg/l) (Sharfa & Subyani, 2011e).

	Max	Min	Mean	St.Dev.	Skew	Co.Var.
As	0.09	0.00026	0.015	0.023	1.98	1.55
Zn	0.36	0.0001	0.044	0.069	3.32	1.59
P	4.16	0.046	0.626	0.540	3.35	0.86
Pb	0.07	0.001	0.020	0.020	1.41	1.00
Ba	0.56	0.037	0.208	0.088	0.64	0.42
B	8.54	0.372	1.612	1.390	2.45	0.86
Si	25.06	4.89	9.832	3.004	2.07	0.31
Al	0.90	0.00256	0.062	0.152	3.25	2.46
Li	0.60	0.002	0.089	0.145	2.35	1.62
Cu	0.06	0.0013	0.021	0.012	0.64	0.58
Fe	0.32	0.005	0.032	0.054	4.42	1.68
Hg	0.17	0	0.018	0.038	3.05	2.10
Mn	0.28	0.001	0.020	0.037	5.10	1.84
Cd	0.02	0.0001	0.004	0.004	1.71	1.06
V	0.70	0.00213	0.213	0.196	0.19	0.92
Rb	0.16	0.0046	0.097	0.055	-0.42	0.57
Ph	7.90	7.1	7.487	0.205	0.36	0.03

Oxidation and dissolution rocks of gabbroic composition will contribute trace elements such as, Cu, Zn, Li, Cd, V and Mn. The source of these trace elements will be associated with the major cations such as Ca^{2+}, Mg^{2+}, Fe, N^{a+}, Al and Si which are considered the major cations in the rocks of gabbroic composition. Whereas, in the weathering granitic rocks will contribute trace elements of Cu, Ba, Rb, Pb and As, These elements will be associated with the major ions such as Na^{+}, K^{+}, Al and Si.

In Wadi Fatimah the groundwater salinity ranged between 969 to 26.500 μS/cm. All the sub- basins are characterized by a relatively low to medium saline water. Highly mineralized water zone almost found marked the lower part of the wadi e.g. Al Jumum–Bahrah area, where the groundwater salinity reached up to 26.500 μS/cm with an average of about 9500 μS/cm.

Based on the irrigation indices (e.g. EC, SAR, RSC, B, MH and chloride hazard), it can be concluded that the groundwater within with exception of those located in the Al Jumum- Bahrah region in Wadi Fatimah basin, although its salinity ranged from medium to high, the groundwater can be safely used for irrigation.

The groundwater in Fatimah basins is suitable for home usage. Analysis conducted within Wadi Fatimah revealed that the majority of the groundwater is highly mineralized and excessively hard, making it unsuitable for use as a drinking water source. Furthermore, certain trace elements had elevated levels that surpassed the permissible thresholds, thereby rendering the water unsuitable for use. These elements include Arsenic (As), Boron (B), and Molybdenum (Mo).

5. CONCISE SUMMARY

The findings of this study indicate that the majority of the water resources in the Jeddah-Makkah area are located in the subterranean aquifers. There are two primary types of aquifers: 1) Shallow underground water sources found in the Quaternary alluvial deposits, and 2) The Oligo-Miocene layers of sedimentary rocks known as Ash Shumaysi and Haddat Ash Sham formations. The exploration instruments used in Wadi Usfan yielded the following conclusions: a) The shallow zone has a saturated thickness ranging from 3-20 meters and water is present under confined to semi-confined conditions. b) Water levels were found at depths ranging from 3-16 meters in the alluvial wadi deposits, and from 18-62 meters in the sedimentary succession. c) Groundwater flows towards the west and northwest, following the overall surface drainage system. d) The hydraulic gradient varies significantly from one location to another, depending on the pumping rates, cross-sectional area of the aquifer, and its transmissivity.

Inside a wadi In a particular region, the salinity of the groundwater is rather low, with electrical conductivity (EC) levels ranging from 542 to 5400 μS/cm, and an average of approximately 1539 μS/cm. The primary water kinds are sodium chloride (NaCl) and calcium chloride (CaCl2). The chemical studies indicate that the groundwater contains high levels of nitrate (NO3) ions and is predominantly characterized by high hardness, rendering it unsuitable for use as a drinking water source. Furthermore, several trace elements exhibited an elevation in their concentrations, above the prescribed upper limits set for their use in drinking water.

The groundwater salinity in Wadi Fatimah varied from 969 to 26,500 μS/cm. A highly mineralized water zone has been identified in the lower section of the Al Jumum–Bahrah area, specifically in the wadi. The salinity of the groundwater in this zone can reach up to 26,500 μS/cm, with an average of approximately 9,500 μS/cm. Considering the irrigation indices such as EC, SAR, RSC, B, MH, and chloride hazard, it can be inferred that the groundwater in the Wadi Fatimah basin, except for the Al Jumum-Bahrah region, is suitable for irrigation despite its moderate to high salinity levels.

REFERENCES

Al Kabir, M. A.(1985) Recharge characteristics of groundwater aquifers in Jeddah-Makkah- Taif area. M.Sc. Thesis, Faculty of Earth Sci., King Abdulaziz Uni., Jeddah, Saudi Arabia.

Alawi, J. and Abdulrazzak, M. 1993. Water in the Arabian Peninsula: Problems and Prospective. In P. Rogers, and P. Lydon (eds), Water in the Arab World Perspectives and Prognoses:171-202. Division of Applied Sciences: Harvard University.

Al-Gamal,S.A., and Sen,Z.(1983) Quantitative analysis of groundwater quality in Western Saudi Arabia. Proc. Int. Symp. Noordwijkuhout.

Al-Hageri, F.Y., 1977. Groundwater studies of Wadi Qudaid. Institute of applied geology, King Abdulaziz University, Jeddah, Saudi Arabia. Research series no. 2. 132- 178.

Al-Khatib,E.A.B.(1977) Hydrogeology of Usfan District. M.Sc. Thesis, Institute of Applied Geology, King Abdulaziz Uni., Jeddah, Saudi Arabia.

Al-Nujaidi,H,A.(1978) Hydrogeology of Wade Murawani- Khulais. M.Sc.Thesis, Institute of Applied Geology, King Abdulaziz Uni., Jeddah, Saudi Arabia.

Alyaamani, MS., and Hussein,M.T.(1995) Hydrochemical study of groundwater in recharge area, Wade Fatimah basin, Saudi Arabia. Geo Journal, 37.1, p:81-89.

Alyamani, M. (2007) Effects of cesspool system on groundwater quality of shallow bedrock aquifers in the recharge area of Wadi Fatimah, Western Arabian Shield, Saudi Arabia. J. Env. Hydrology, vol. 15, paper 8.

Alyamani, M.S.(1999). Physio-chemical processes on groundwater chemistry, under arid climatic conditions,western province, Saudi Arabia. Proj. No. 203/418, King Abdulaziz Uni., Jeddah, Saudi Arabia.

Alyamani, M.S., Bazuhair, A. S., Bayumi, T. H., and Al-Sulaiman, K. (1996) Application of environmental isotope on groundwater study in the western province, Saudi Arabia. Proj. No. 005/413, King Abdulaziz Uni., Jeddah, Saudi Arabia.

Basmci, Y., and Al Kabir, M.A. (1988). Recharge characteristics of aquifers of Jeddah-Makkah-Taif region. Mathematical and Physical Sci., Vol. 222, p: 367-375.

Bazuhair,A.S.,Hussein,M.T.,Alyamani,M.S.,and Ibrahim,K.(1992) Hydrogeophysical studies of Khulais basin, western region, Saudi Arabia. Unpub. Report, King Abdulaziz Uni, Jeddah, Saudi Arabia.

Bolt, G. H. and Bruggenwert, M. G. M.: 1978, Soil Chemistry, Basic Elements, 2nd Edition, Elsevier Scientific Publishing Company Amsterdam. The Netherlands.

Brady, N. C.: 1974, Nature and properties of soils, 8th EditionMacmillan Publishers Co., INC, NewYork Brady, N. C.: 1974, Nature and properties of soils, 8th EditionMacmillan Publishers Co., INC, NewYork.

Brown, G.F., Jackson, R.O., Bogue, R.G., and MacLean, W.H. (1963) Geology of the Southern Hijaz quadrangle, Kingdom of Saudi Arabia: Saudi Arabian Dir. Gen. Min. Res.Misc. Geologic Invest. Map I - 210A, 1:500,000 scale.

Burden, D.J. 1982. Hydrogeological conditions in the Middle East. Quarterly Journal of Engineering Geology 15: 71–81.

Edmunds. W. M., Kinniburgh, D. G. and Moss, P. D.: 1992, Trace Metals in Interstitial Waters from Sandstones Acidic Inputs to Shallow Groundwater, Environmental Pollution, Elsevier Science Publishers Ltd, England. Edmunds, W. M. and Savage, D.: Geochemical Characteristics of Groundwater in Granites and Related Crystalline Rocks. (Unpublished).

Hem, 1989 Hem, J.D. (1989) The Study and Interpretation of the Chemical Characteristics of Natural Water. 3rd edn. USGS Water Supply Paper 2254, US Geological Survey.

Italconsult (1976) Detailed investigations of Wade Khulais basin. Unpub. Report. Ministry of Agriculture and Water, Riyadh, Saudi Arabia

Jamman, A. M. (1978) Hydrogeology of Wadi An Numan, Saudi Arabia. Unpub M. Sc.Thesis, Faculty of Earth Sci., King Abdulaziz Uni., Jeddah, Saudi Arabia. M.Sc. Thesis, Institute of Applied Geology, King Abdulaziz Uni., Jeddah, Saudi Arabia.

Johnson, P. R. (2006) Explanatory notes to the map of Proterozoic geology of western Saudi Arabia, technical report SGS-Tr-2006-4.

Karpoff, R.,1958. Esquisse geologiques de I,Arabie Saoudite : Bulletin Societe Geologique de France, 6 ser. VIIm, pp. 653 – 696.

Kfupm/Ri (King Fahd University of Petroleum and Minerals, Research Institute) 1988. Groundwater Resources Conditions in the Eastern Province of Saudi Arabia, Research report, King Fahd University of Petroleum and Minerals, Research Institute. 61 pp

Liptrot, G. F.: 1989, Modern Inorganic Chemistry 4th Edit, Scoltprint Musselburgh.

Lloyd and Heathcote, 1985 Lloyd, J. A., and Heathcote, J.A. (1985) Natural inorganic hydrochemistry in relation to groundwater: An introduction. Oxford Uni. Press, New YorK p: 296.

Mansour, M. A. (1984) Evaluation of groundwater resources of Wade Fatimah by Numerical model. Unpub. M.Sc.Thesis, Faculty of Earth Sci., King Abdulaziz Uni., Jeddah, Saudi Arabia.

MAW (Ministry of Agriculture and Water) 1984. Water Atlas of Saudi Arabia. Department of Water Resources Development, Ministry of Agriculture and Water, Riyadh, Saudi Arabia.

Moore, T. A., and Al-Rehaili, M.H. (1989) Explantory notes to the geologic map of the Makkah quadrangle, sheet21D, Kingdom of Saudi Arabia: Saudi Arabian Dir. Gen. Min. Res. Geoscience map GM-107C, 1:250,000 scale.

Sharaf, M. A. M. (2010a): Geophysical and Hydrochemical Studies on the Groundwater Aquifer in Ash Shamiyah Area, Makkah District, Western Saudi Arabia. Journal of Environmental Hydrology, Paper 13, V. 18, 2010).

Sharaf, M. A. M. (2010b): Hydrogeology of Haddat Ash Sham area: Geophysical and hydrochemical Constrains. Fifth 5th International Conference on the Geology of the Tethys Realm, South Valley University, January 2010, P.147-158.

Sharaf, M. A. M. (2011a): Hydrogeological and hydrochemistry of the aquifer system of Wadi An Numan, Makkah Al Mukarramah, Saudi Arabia. AquaMundi: Journal of Water Science, (2011)-Am03027: 035 – 052

Sharaf, M. A. M. (2011b): Geological and Geophysical Exploration of the Groundwater Aquifers of As Suqah Area, Makkah District, Western Arabian Shield. Arab J Geosci (2011) 4:993–1004.

Sharaf, M. A. M. (2011c): Hydrochemistry of the Groundwater Aquifer in wadi Faydah- Usfan Areas, Makkah District, Western Arabian Shield, Saudi Arabia. Egyptian Journal of Geology, v. 55, 2011, p.1-13.

Sharaf, M. A. M. (2011d): Hydrochemistry of the Groundwater Aquifer in As Suqah Area, Makkah District, Western Arabian Shield, Saudi Arabia. (In Press in the Journal of King Abdulaziz University (Earth Sciences) JKAU, V. 22, 2011).

Sharaf, M.A.M and Subyani, A. M. (2011e): Assessing of Groundwater Contamination by Toxic Elements through Multivariate Statistics and Spatial Interpolation, Wadi Fatimah, Western Arabian Shield, Saudi Arabia. International Journal of Scientific & Engineering Research Volume 2, Issue 9, September-2011.

Sharaf, M. A., Farag, M. H., and Gazzaz, M. (1988) Groundwater chemistry of WadiUoranah- Alabdiah area. Western Province, Saudi Arabia, JKAU: Earth Sci. 1, 103-112.

Sharaf, M.A., Al-Bassam,A., Bayumi, T.H., and Qari,M. H. (2002) Hydrogeological and hydrochemical investigation of the Cretaceous- Quaternary sedimentary sequence east of Jeddah city. King Abdulaziz City for Science and Technology (KACST), Final Report, Riyadh, Saudi Arabia.

Spincer, C. H.,and Vincent, P. L., 1984, Bentonite resource potential and geology of the Cenozoic sediments, Jeddah region: Saudi Arabian Deputy Ministry for Mineral Resources, Open-File Report BRGM-O-F-02-34, 34 p.

USSL 1954: Diagnosis and improvement of saline and alkali soils. USDA Agr. Handbook No. 60, Washington DC WHO (1993) Guidelines for Drinking Water Quality, Geneva

Zaidi, S. (1983) Landform and geomorphic evolution of Wade Khulais area.Western Saudi Arabia. FES, Bull. 5, 153-156pp.

Zaidi, S.(1984) Geomorphology of Wadi Khulais area. FES, Res. Ser. No.18, 98p.

07 APPLICATION OF REMOTE SENSING AND GEOGRAPHIC INFORMATION SYSTEM (GIS) FOR THE ANALYSIS OF DRAINAGE SYSTEM IN THE OCCURRENCE OF FLOODS IN JEDDAH - WESTERN SAUDI COAST

1. INTRODUCTION: AN OVERVIEW

Recently, there has been a rise in natural disasters, which has garnered significant attention both regionally and globally. Flooding is a highly severe manifestation of these dangers, resulting in the deaths of over one million individuals annually in low-income countries, according to the International Disaster Database (EM-DAT: The OFDA/CRED, 2010).

Saudi Arabia, situated on the Arabian Peninsula, occupies a significant portion of the region. It is positioned in an active tectonic zone known as the Red Sea Rift System, resulting in an unstable area with complex geology and geomorphology. Consequently, the country's land has seen numerous natural disasters throughout its history, primarily centered in the western coast between the Arabian Shield and Red Sea, which is where the area of study is located. Various natural dangers have been observed in this region, mostly include floods, earthquakes, and dust storms. Inhabitants typically prioritize the consideration of these hazards due to their direct influence on their lifestyle, while other factors such as soil and rock erosion and drought are often overlooked.

However, after a significant period of time, this region had not had any notable catastrophic occurrence until the late 2009, when there was heavy rainfall in the city of Jeddah and the surrounding areas, resulting in destructive floods. This city is situated in the central region of the western coast of Saudi Arabia. It is recognized as the primary commercial and tourist hub of the country, with a growing population of over 3.4 million individuals. Additionally, it serves as the primary intersection for pilgrims hailing from many nations across the globe.

According to the statistical overview provided by the International Disaster Database (EM-DAT: The OFDA/CRED, 2010), Saudi Arabia saw a total of 299 fatalities due to various natural disasters between the years 1982 and 2005. This, in turn, demonstrates the extent of the harm caused by the flood disaster that happened in 2009. This demonstrates the extent of the harm caused by the most recent flood disaster. In January 2011, a subsequent flood occurred, affecting a wider area compared to the previous year. Furthermore, this incident caused significant destruction and resulted in multiple fatalities, losses, and injuries.

Research on this topic was few and part of it solely addressed technical matters, making it challenging to fully comprehend the causes and mechanisms of catastrophic catastrophes and disaster response. Consequently, without thorough studies that specifically examine the behavior and characteristics of the current drainage system, it is not possible to effectively implement measures to reduce risk and mitigate potential problems.

Fig. 1. Location maps of the Saudi Kingdom and Jeddah city.

The study seeks to examine the topological components of drainage systems and deduce their geomorphological and hydrological features. The display exhibits the geospatial data obtained from satellite imagery and the utilization of Geographic Information Systems (GIS). Digital Elevation Models (DEM) were also utilized to evaluate significant factors for drainage study.

The study region was chosen based on the correlation between the boundaries of the current water basins and the surrounding areas that are prone to recurrent flooding, as evidenced by the floods in 2009 and 2011 (Figure 1). Hence, the resultant expanse measures approximately 1947 square kilometers and spans from the shoreline to the neighboring mountain ranges. The area is situated within the geographical coordinates of 39°32′ and 39°06′ E longitude, and 21°56′ and 21°1′9N latitude.

2. NATURAL AND ANTHROPOGENIC CHARACTERISTICS

2.1 Drainage systems

Advanced techniques have been well demonstrated in illustrating streams and surface water basins. In this respect, DEM can be well utilized, and it can help inducing flow directions, and locating low-lands. However, erroneous results sometimes appear from DEM if the digital applications do not accurately obtain. Topographic maps are also used to delineate drainage systems. Therefore, streams are directly digitized in the GIS system, and thus catchment areas are extracted following geomorphologic and hydrological concepts.

Table 1. Watersheds in the area of study and their types.

Basin No.	Basin	Area (km^2)	Basin type	Basin No	Basin	Area (km^2)	Basin type
1	Ghouimer	319.7	Major	16	Selsli	13.7	Joining
2	Om El- Hableen	75.7	Major	17	Muwaieha	23.7	Minor
3	Basin # 3	6.5	Joining	18	Basin # 18	17.6	Joining
4	Daghbj	56.9	Major	19	Basin # 19	14.8	Minor
5	El Hatiel	59.6	Major	20	Basin # 20	21.6	Joining
6	Basin # 6	10.3	Joining	21	Abou Je'Alah	21.2	Minor
7	Basin # 7	25.8	Minor	22	Al A'ayah	14.7	Minor
8	El Assla	289.4	Major	23	Ed-Dowikhlah	17.1	Minor
9	Basin # 9	10.4	Joining	24	El-Baghdadi	29.6	Minor
10	Mreikh	46.7	Minor	25	Ketanah	34.7	Major
11	Kawes	70.1	Major	26	Basin # 26	13.0	Joining
12	Osheer	17.7	Minor	27	Esh-Shoabaa	40.5	Major
13	Basin # 13	12.5	Joining	28	Basin # 28	24.4	Joining
14	Methweb	54.2	Major	29	Da'af	37.9	Major
15	Ghlil	23.1	Minor				

In this study, topographic maps, of scale 1:50000 and contour interval 20m, were utilized and they were supported by the application of DEM in order to extract the related parameters for drainage systems, and they were directly digitized in the GIS system using *Arc-GIS 9.3* software. Consequently, streams were illustrated for each watershed and the watershed boundaries were identified. The area of study is about 1947km2 (Figure 2 and Table 1).

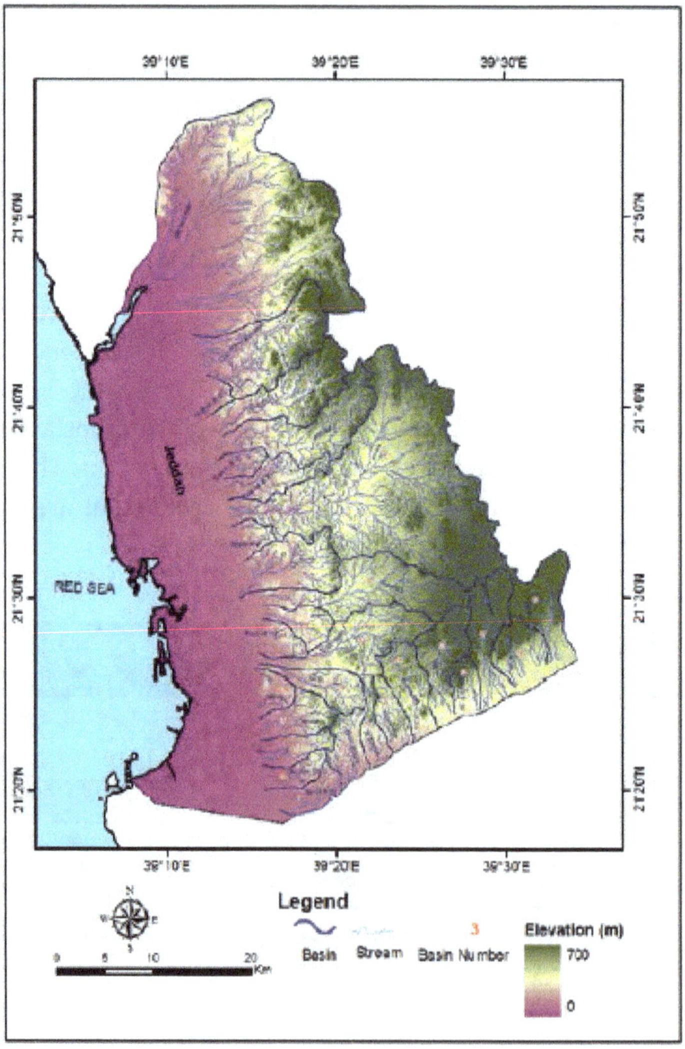

Fig. 2. Watersheds in the area of study.

Accordingly, the watersheds in the area of concern were divided into three types as follows:

1. Major basin: It encompasses principal hydrological characteristic, mainly funnel-like shape, where the difference between the numbers of branches is high in upstream and downstream areas, and it is characterized by uniform run-off. There are 10 major basins in the study area.

2. Minor basin: It is characterized by smaller areas than the major ones (usually less than 50km2), and it has almost a uniform run-off regime. They are 10 minor basins existing in the study area.
3. Joining basin: It is a geographic land extension between the major and minor basins. It is characterized by non-uniform run-off. There are 9 joining basins in the study area. There are 29 watersheds in the area of study, 19 of them outlet towards Jeddah city at the coast, and totaling an area of about 1170 km2, while the rest outlet into Wadi Fatima to the southern side of Jeddah city. The largest watersheds are Ghouimer (319.7km2) and El Assla (289.4 km2).

2.2 Rainfall distribution

There is still undefined annual rainfall rate for Jeddah area. It was estimated at 350-400mm in the last few decades (Italconsult, 1967), while it has been recently estimated with less than 60mm, whereas, run-off is estimated between 5-6% according to Es-Saeed, et al. (2004). This decline in annual rainfall rate (i.e. about five times) is accompanied with a number of rainfall peaks, which is attributed mainly to the changing climatic conditions (IPCC, 2007).

Due to the lack of continuous and comprehensive climatic data, the remotely sensed data, which can be retrieved from satellite images, was used. Therefore, Tropical Rainfall Mapping Mission (TRMM) data was utilized in this study, and thus the geographic distribution, through maps, of rainfall in November 2009 and January 2011 were illustrated from TRMM data (Figure 3).

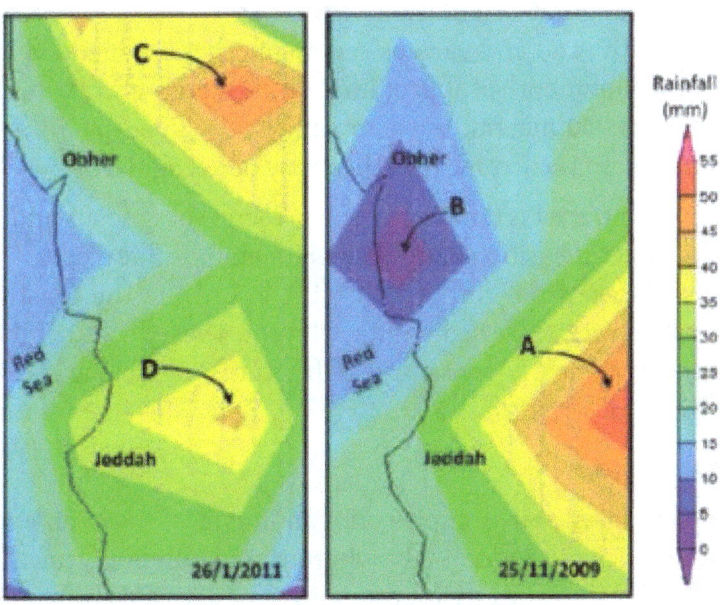

Fig. 3. Rainfall distribution in 2009 and 2011 according to TRMM data.

These maps could introduce essential tools to figure out the geographic distribution of rainfall in the two periods, which will be used consequently to compare it with the geographic distribution of damaged areas. Hence, it is clear that there is one dense cloudy domain in each event (2009 and 2011). These are A in 2009 and C in 2011, in addition to other less-dense ones. However, this was not very well realized previously. For the dense cloudy domain in 2009, it was estimated with a diameter of about 90km, whilst the other one of 2011, it was about 30km, and this reflects the degree of impact in the two periods. However the rainfall rate was estimated at about 95mm, and 120mm in 2009 and 2011; respectively; but this was attributed to a gauge point and not for the entire geographic distribution of rainfall.

2.3 Features related to the shape and structure of the Earth's surface and the study of rocks and their formation

The study of floods relies heavily on the analysis of geomorphologic and geologic properties, since they determine the flow regime and several related surface processes such as infiltration rate, flow direction, and water accumulation. The field of research can be categorized into three primary regions based on their elevation and the prevailing rock formations. The following items are:

1. Coastal area (10-20km width, 3m/km slope gradient).
2. Slopping lands (<5km width, 20m/km slope gradient).
3. Elevated areas (about 25km width, 6-7m/km slope gradient).

Furthermore, the study region encompasses a significant quantity of valleys, exhibiting a wide range of orientations and characteristics. The containers are nearly saturated with sand and various sediments that are blended together. The valleys exhibit a shallow depth of approximately 2-3 meters and a wide cross-section ranging from 500 to 700 meters. In these valleys, the flood plains are frequently indistinct or shallow to the extent that they may be recognized.

The geology of the area is defined by the presence of igneous and sedimentary rocks that are sometimes interlayered with metamorphic rocks. The majority of the visible rocks are from the Precambrian era, while the coastline region is completely overlaid with loose sedimentary layers. These rocks exhibit significant deformation systems, mostly consisting of fractures of varying sizes, joints, and various folding characteristics.

2.4 Urban expansion

The kingdom of Saudi Arabia occupies outstanding level in urban expansion, and this is well pronounced in Jeddah city where the population growth rate ranges between 20-28%, and thus the city is occupying now about 3.4 million people (3500 person/km2). This is merely attributed to the number of foreign people attend during different periods for pilgrims and related religious tasks.

Lately, the geographic distribution of urban settlements has been extended towards the mountainous areas to the east. However, this distribution extends along the existing valleys from the elevated regions, which results obstacles in many valleys passageways. Accordingly, the interpretation of satellite images in combination with the analysis of old topographic maps show that urban expansion has been three times increased over 21 years (1975-1996), and thus it is followed by another increase of about 2.7 time in the last fifteen years (1996-2011), and this increase concentrates mainly along the eastern part of Jeddah city.

3. MATERIALS AND METHODS

There are many methods to study and assess floods and their controlling geomorphologic factors. They usually follow different approaches of analysis, and thus different results exist. This relies on the used tools; however, the utilization of new space techniques is utmost important in this respect, since they became of essential role for topological and drainage analysis. However, the use of such tools is still dependant on its availability. In addition, many studies focus on specific concept relates to floods and torrents, such as the analysis of hydrologic systems (Subyani, 2009), Digital Elevation Models (KACST, 2011), or the morphometric analysis for valleys (Yehia and El-Ater, 1997).

3.1 Used tools

3.1.1 Information and data

- Climate data from ground stations to induce the frequency of rainfall peaks by region.
- Climate data from space sources to cover the lack of ground data.
- General ground data (e.g. damaged areas, specifications of the channels, etc.).
- Supplementary information, including historical records of floods, urban expansion, etc

3.1.2 Maps

- Topographic maps of 1:50.000 scale, with contour interval 20 meters.
- Geological maps of 1:250.000 and 1:500.000 scale.

3.1.3 Satellite images

Satellite images of different optical and spectral specifications were used in this study. They have different characteristics, such as re-visit time, swath width, etc. there was a great concern to have these images in dates close to the disasters dates of disasters (Table 2).

Table 2. Used satellite images and their major specifications.

Satellite	Spatial resolution	No. of bands	Acquiring date	Utility
Ikonos	1m	5	10/10/2009, 30/11/2009, 19/2/2010	Comparison before and after floods. Identification areas under flood damages (2009).
Worldview-1	0.5m	8	27/1/2011, 1/3/2011	Identification areas under flood damages (2011). Spatial data overlapping.
Worldview-2	0.5m	8	8/2/2011	
Quick Bird	0.5m	5	2/3/2011	High precision of comparative analysis for flooded areas.
Geo-eye 1	0.5m	5	2010	Establishing DEM
Aster	15m	14	2009	Identification of geological features.

3.1.4 Digital Elevation Model (DEM)

Digital Elevation Model (DEM) was analyzed in this study to induce empirical approaches for drainage analysis. This DEM data was supported by King Abdulaziz City for Science and Technology.

- Data for Digital Elevation Model (DEM), with 2 meters precision.
- Data for Digital Elevation Model (DEM), with 30 meters precision.

3.1.5 Software

- *ENVI-4.3* for satellite image processing, produced by *IBM*, Colorado, USA.
- *ERDAS Imagine 9.3* for satellite image processing, produced by *Leica*, Georgia, USA.
- *Arc-GIS 9.3* for GIS applications, produced by *ESRI*, Redlands, USA.

3.2 Methods of analysis

3.2.1 Images processing

Satellite images have become important tools of Earth observation since they can be used in several terrain applications and the processes occur. The science of remote sensing, in a broad sense, is represented by digital satellite images acquired in defined technical procedures. All selected digital satellite image data were primarily subjected to two principal stages: these are the: pre-processing and images processing procedures.

Pre-processing procedures are essential and diverse set of image preparation programs that act to offset problems. These processes are applied into the specialized software (e.g. *ERDAS Imagine, ENVI, etc.*) to increase the accuracy and interpretability of the digital data during the image processing phase. These commonly include: image sub-setting, atmospheric correction, geometric correction, image registration, geo-referencing and mosaicing.

Consequently, digital image processing can be applied to facilitate recognition of objects appear on terrain surface, and thus applying different analyses and measures required. For example, in *ERDAS Imagine* software, the most useful steps are: directional filtering, contrasting and sharpness. In addition, band combination is also applied, where single band and multi-band enhancement were carried out by interrelating each three bands as one set. These applications provided helped detecting color differentiation, pattern and tone, which would discriminate distinguished geomorphologic features. Moreover, thermal interpretation from thermal bands was applied. While *ENVI* software is also useful and can be friendly used for digital satellite images processing. The following digital applications were performed: Enhancement, Interactive stretching, Density slicing and coloring.

3.2.2 GIS applications

The capability of using GIS technology for earth surface based scientific investigation makes it more usable; especially for geographic applications when satellite images are not available. Modern GIS technologies use digital information, for which various digitized data creation methods are used. The most common method of data creation is digitization, where a hard copy map or survey plan is transferred into a digital medium through the use of a computer-aided design and geo-referencing capabilities.

Arc GIS 9.3, which was used in this study, is a software program used to create, display and analyze geospatial data. It consists of three components: Arc Map, Arc Catalog and Arc Toolbox. Arc Map is used for visualizing spatial data, performing spatial analysis and creating maps to show the work results. While, Arc Catalog is used for browsing and exploring spatial data, as well as viewing a creating metadata and managing spatial data. Arc Toolbox is an interface for accessing the data conversion and analysis function the come from *Arc GIS*.

4. DATA ANALYSIS AND RESULTS

4.1 Three-dimensional models (DEM) production

Geographic information systems can be used also to build three-dimensional models for any geographical location on the surface of the Earth. The representation of terrain topography of any site needs data form three-dimensions (z, y, x), which is known as digital elevation model (DEM). Thus, the applications of DEM is well pronounced in several applications, and more certainly to induce many components, such as slope, sunlight exposure, drainage systems, low-lands, etc.

The concept behind establishing DEMs implies the treatment of elevation points whether from digital contour lines or from stereoscopic satellite images (e.g. SPOT images). Hence, triangulated irregulated network *(TIN)* must be primarily constructed, which represents digital data structure used in *GIS* of surface attributes of the physical land surface or sea bottom, made up of irregularly distributed nodes and lines with three dimensional coordinates (*X*, *Y* and *Z*) that are arranged in a network of non-overlapping triangles (Figure 4). *TIN*s are madefrom mass points, break-lines, and polygons. Mass points are height points; they become nodes in the network, thus they primary input into a *TIN* in order to determine the overall shape of the surface.

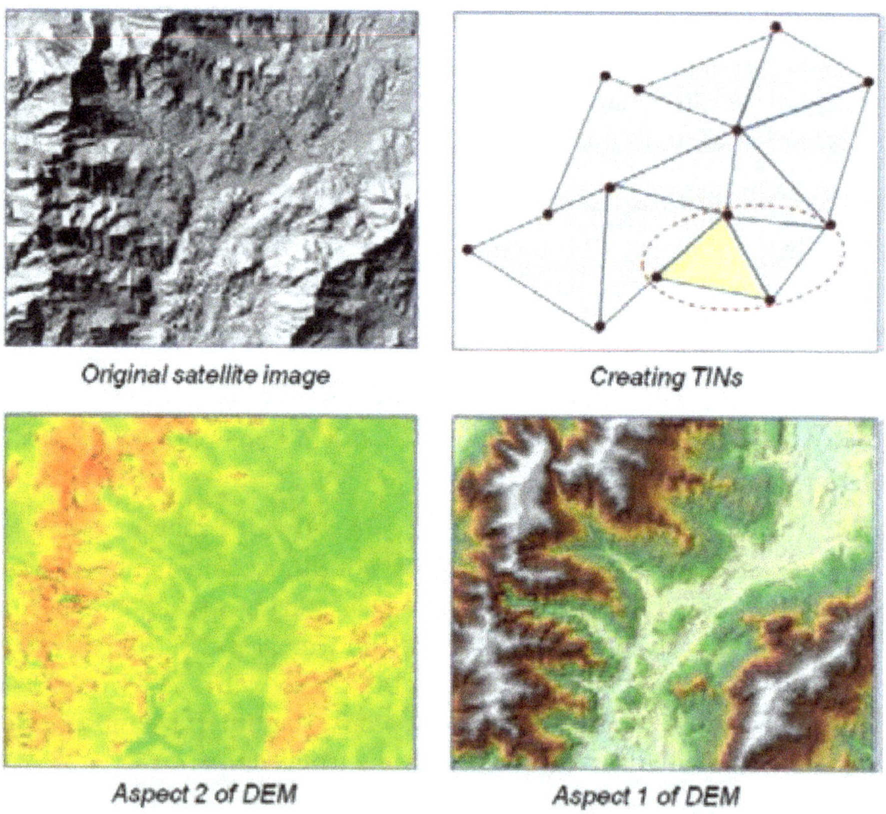

Fig. 4. Example showing the steps obtained to create DME aspects.

4.2 Analyzing digital elevation mode

Currently, the use of Digital Elevation Models (DEM) has found widespread applications in several geomorphologic and hydrological purposes; especially that this technique allows the extraction of topographic features of the earth surface making possible the display of all natural features for the both vertical and horizontal resolutions. However, DEM extraction requires elevation data for topographic features with geographic coordinates of each elevation. Therefore, topographic maps with contour lines are electronically digitized in specialized GIS software. In case these maps weren't available at a digital format, satellite images with stereoscopic characteristics can be used. In this study, Geo-eye-1 satellite images, with 2m precision were used.

In this study, after establishing the DEM, the resulting data was analyzed to recognize slope gradient, cross-sections, channel slopes and depressions. Hence, the three first factors interact together to form water flow network from drainage system thresholds to the valleys outlet.

4.2.1 Terrain slope

Terrain slope is considered as an important factor in identifying the flood regime. In this study, terrain slopes are considered rather than the valley slopes. Naturally flood risk increases with terrain slope gradient (angle of inclination with the horizontal) due to the increasing energy of run-off resulted from the surrounded surfaces to the adjacent valleys. It is viewed from the capacity of valleys receiving water bulk and bed loads from these surfaces, which may result flooding.

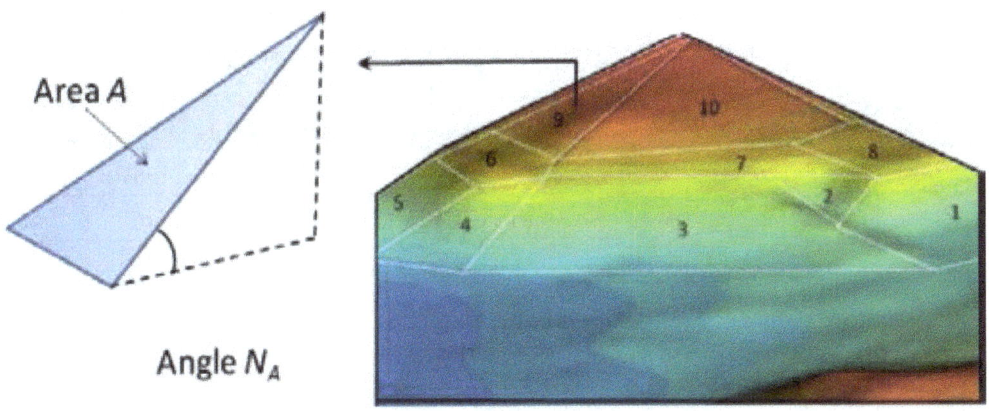

Fig. 5. Classifying terrain surfaces according their slopes for each surface.

In order to evaluate these geometric variables from DEM, every surface must be taken independently (i.e. terrain surface in certain exposure), and since these variables will be used in studying the overland flow, it was necessary to analyze the variation required for every water basin separately in the area of study. Since water basins consist of a set of surfaces expanding toward valleys; however, it can be dividing each basin into zones with a specified surfaces slopes and defined area for each surface. It makes it possible to evaluate the total slope effectiveness in every basin in the study area. This can be done by creating DEM in GIS; and more specifically in Arc-GIS software (example in Figure 5).

Nevertheless, slope gradient of the surface is taken from a linear approach where it is described as the length slope (Ls) according to several studies (Khosrowpanah, et al., 2007), and the total surface of the area is not considered.

In order to evaluate the effectiveness of the total slope in the surface water flow energy, the area (basin) must be divided into a number of surfaces, where each surface is characterized by a defined slope angle (level of inclination) (Na). Therefore, the areas of these surfaces (A) with their angles will be calculated.

Thus, the exponential interaction of to surface inclination with its area (multiplication calculation), can induce the effectiveness (St) of water volume loaded towards valleys. It can be calculated to every surface then the sum can be calculated by dividing the resulted values on the number of surfaces (Ns) as follows:

$$St = \frac{(A_1 \times N_{a1}) + (A_2 \times N_{a2}) + (A_3 \times N_{a3}) + \ldots}{N_s}$$

In this study, the surface slope angles were classified into nine intervals ranging between 0º- 10º for the very low slope effectiveness to 80º -90º for the very high slope effectiveness. Thus, slope effectiveness values were calculated for the 29 basins exist in the area of study (Table 3).

Table 3. Slope effectiveness in water basins after calculating values from DEM.

Basin No.	Basin name	(Na)	(A) Km2	Average slope effectiveness (º)	Total effectiveness (St)
1	Ghouimer	23	12.5	°8	Medium
2	Om El-Hableen	16	0.84	10°	Medium
3	Basin # 3	2	0.48	3°	Low
4	Daghbj	20	1.02	15°	High
5	El Hatiel	17	0.96	11°	High
6	Basin # 6	3	0.65	4°	Low
7	Basin # 7	5	2.74	12°	High
8	El Assla	32	5.72	7°	Medium
9	Basin # 9	3	1.43	4°	Low
10	Mreikh	9	1.88	7°	Medium
11	Kawes	18	2.12	9°	Medium
12	Osheer	5	0.98	5°	Low
13	Basin # 13	2	1.97	4°	Low
14	Methweb	13	2.05	11°	High

Basin No.	Basin name	(Na)	(A) Km2	Average slope effectiveness (º)	Total effectiveness (St)
15	Ghlil	7	0.78	4°	Low
16	Selsli	2	0.27	1°	Low
17	Muwaieha	5	1.57	8°	Medium
18	Basin # 18	4	1.18	6°	Medium
19	Basin # 19	3	0.87	2°	Low
20	Basin # 20	1	0.58	1°	Low
21	Abou Je'Alah	5	1.08	9°	Medium
22	Al A'ayah	4	1.49	7°	Medium
23	Ed-Dowikhlah	8	0.97	5°	Low
24	El-Baghdadi	17	1.07	12°	High
25	Ketanah	19	1.24	14°	High
26	Basin # 26	7	0.58	15°	High
27	Esh-Shoabaa	19	1.22	17°	High
28	Basin # 28	9	1.13	13°	High
29	Da'af	14	1.76	15°	High

4.2.2 Cross sections

Cross sections of valleys are commonly used in many geomorphologic and hydrologic studies that rely on evaluating tributaries and the mechanism of run-off of water and the bed load, whether for run-off rate or erosion and sedimentation. It is well known that the characteristics of cross-section of tributaries separately have no impact on the hydrological system for flood assessment, but it needs an interaction with other hydrological and geomorphologic characteristics.

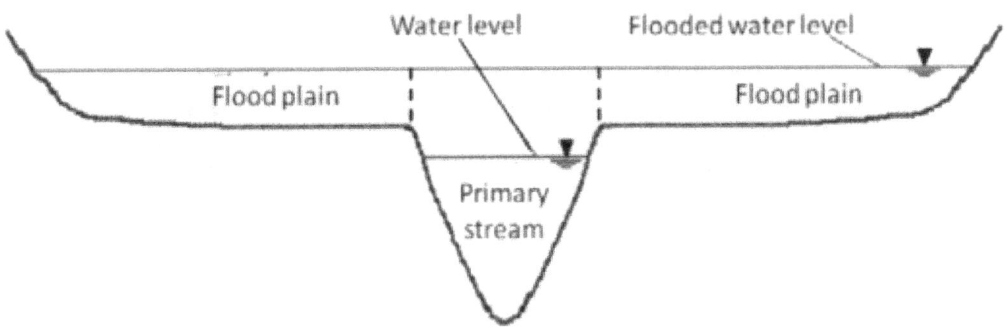

Fig. 6. Schematic cross-section showing major dimensions of a valley.

Cross section variables to any water course comprise the width, depth and the length of the primary stream, but in the case of wide flood plain, with shallow depth (as in the study area); however, width and depth of these plains will be included too. In the case of wide and shallow flood plains, run-off water can easily overflow into these plains when water level increases in the primary watercourse even with a small amount because the difference in depth, of the primary watercourse and flood plains, is relatively negligible (Figure 6).

Normally, the width and depth of water channel are not constant, and they vary with valley length. Therefore, in calculating the capacity of each channel, it must be divided into different channel segments each with specific depth, width and length. Where the length is determined according to channel diversions (i.e. a section is attributed to each stream segment with specific direction). When the three geometric variables (length, width and depth) are determined, with those variables of the flood plain, thus the total section area of each segment can be calculated and the sum of the resulted values will extend the total cross-section area of the valley (Xt). It can be calculated as follows:

$$X_t = \frac{\{[(W_{m1} \times D_{m1}) + (W_{f1} \times D_{f1})] \times L_1\} + \{[(W_{m2} \times D_{m2}) + (W_{f2} \times D_{f2})] \times L_2\} + \ldots}{N_{sg}}$$

- W_{m1} is the width of the main valley, No. 1.
- D_{m1} is the depth of the main valley, No. 1,
- W_{f1} is the width of flood plains No.1,
- D_{f1} is the depth of flood plain No.1,
- L_1 is the segment length,
- N_{sg} is the number of segments.

It is possible to calculate geometric variables using DEM if it is characterized with high resolution that fits with the width and depth of cross-sections. The analysis is done by dividing the sections into different segments. Thus, Arc-GIS software is used to calculate the variables to each segment (Table 4).

Table 4. Cross-section areas of primary valleys as calculated from DEM.

Basin No.	Basin name	No. of Channels	(Xt) Km2	(St)
1	Ghouimer	4	0.126	Medium
2	Om El-Hableen	2	0.215	High
3	Basin # 3	1	0.095	Low
4	Daghbj	2	0.128	Medium
5	El Hatiel	2	0.075	Low
6	Basin # 6	1	0.102	Medium
7	Basin # 7	1	0.087	Low

Basin No.	Basin name	No. of Channels	(Xt) Km^2	(St)
8	El Assla	8	0.121	Medium
9	Basin # 9	1	0.230	High
10	Mreikh	1	0.183	High
11	Kawes	4	0.182	High
12	Osheer	1	0.105	Medium
13	Basin # 13	1	0.021	Low
14	Methweb	1	0.099	Low
15	Ghlil	1	0.258	High
16	Selsli	1	0.177	High
17	Muwaieha	1	0.183	Low
18	Basin # 18	1	0.172	High
19	Basin # 19	1	0.114	Medium
20	Basin # 20	1	0.247	High
21	Abou Je'Alah	1	0.084	Low
22	Al A'ayah	1	0.097	Low
23	Ed-Dowikhlah	1	0.073	Low
24	El-Baghdadi	1	0.035	Low
25	Ketanah	1	0.045	Low
26	Basin # 26	1	0.046	Low
27	Esh-Shoabaa	1	0.076	Low
28	Basin # 28	1	0.104	Medium
29	Da'af	1	0.062	Low

4.2.3 Channel slope

In addition to terrain surface slope that adjacent to valleys; yet the slope gradient (degree of inclination) of the valley channel itself is an important geometric element that must be considered in assessing floods and torrents. Hence, flow rate increases with the degree of inclination of the valley channel. In this respect, flow rate plays a major role in accelerating the process of channel discharge, which is also accompanied with high rate erosion process. Whilst, slow flow rate decreases the discharge rate and consequently influences flood occurrence.

The channel slope can be calculated using the following simplified equation:

$$\frac{L = \text{Lenghth of the channel}}{\Delta h = \text{Difference in elevation}}$$

However it is obvious that stream channels are characterized by different elevations along each valley, and they are not similar. For this reason, sometime it is referred to the following equation introduced by Morisawa (1976).

= (E 0.85 L) - (E 0.10L)/ E 0.75L

Where:

E is the elevation,

= (E 0.85 L) – (E 0.10L)/ E 0.75L

L is a lenght point along the channel,

As for example, (E 0.85 L) means the elevation at 85% distance from the upstream.

The availability of DEM easily enables calculating the channel slope with considerable precision, where the slope gradient is calculated on several defined parts along the valley with more accuracy then using the mathematical equation mentioned above.

Accordingly, in this study, this method was followed and the results are as shown in Table 5. Results show the channel slopes of primary valleys. Thus, having the three main variables (terrain slope, cross sections and their slopes) enables assessing the flow capacity of these valleys, which together influence flood occurrence. Therefore, it is obvious that there are basins more prone to floods and torrents (such as Om El-Hableen and Daghbj basins) than other basins in the area (such as basins no. 1 and 6).

4.2.4 Depressions

Depressions are considered as one the most important terrain features in floods and torrents occurrence. They play a double role in surface water flow and surface water storage, since depressions are naturally retarding water flow and store almost capture water within the terrain surface. As well as, depressions may accumulate water directly from rainfall. Therefore, they are considered as flooded areas, but with least damage.

Hence, identifying depressions directly from satellite images in dry seasons is not an easy task, unless they filled with water for better observations. However, DEM can help identifying depression in any condition, since it relies on elevation differentiation (e.g. 2 meters accuracy). Therefore, depressions characterize terrain surface and enable to cartography the water accumulation from torrential rainfall.

In this study, a new concept was followed in using DEM manipulation to localize depressions.. This concept relies on creating reference lines to the earth surface at defined elevations according to the terrain characteristics. Hence, any elevation below will be considered as a zone with depression.

Table 5. Valleys' slope and their water-bearing capacity.

Basin No.	Basin name	Number of terrain surfaces	Channels No.	General slope Average	(St)
1	Ghouimer	23	4	16	Medium
2	Om El-Hableen	16	2	18	Medium
3	Basin # 3	2	1	23	High
4	Daghbj	20	2	28	High
5	El Hatiel	17	2	19	Medium
6	Basin # 6	3	1	7	Low
7	Basin # 7	5	1	13	Medium
8	El Assla	32	8	16	Medium
9	Basin # 9	3	1	6	Low
10	Mreikh	9	1	13	Medium
11	Kawes	18	4	19	Medium
12	Osheer	5	1	11	Medium
13	Basin # 13	2	1	8	Low
14	Methweb	13	1	16	Medium
15	Ghlil	7	1	14	Medium
16	Selsli	2	1	4	Low
17	Muwaieha	5	1	7	Low
18	Basin # 18	4	1	10	Low
19	Basin # 19	3	1	9	Low
20	Basin # 20	1	1	6	Low
21	Abou Je'Alah	5	1	17	Medium
22	Al A'ayah	4	1	19	Medium
23	Ed-Dowikhlah	8	1	18	Medium
24	El-Baghdadi	17	1	27	High
25	Ketanah	19	1	34	High
26	Basin # 26	7	1	33	High
27	Esh-Shoabaa	19	1	36	High
28	Basin # 28	9	1	18	Medium
29	Da'af	14	1	29	High

Table 6 shows the total area of depression among each water basin. But it must be taken into account that these depressions are not only natural ones, since a large part of the study area is occupied by low-lands that collect water due to excavation and construction (planned zones, roadsides, etc.). In this respect, the integration of satellite images taken after the occurrence of torrential rainfall in the 25th of November 2009 and the 26th of January 2011, with DEM can help in identifying natural depressions from man-made ones.

Table 6. Areas of depressions in different watersheds.

Basin No.	Basin name	Total depressions area (km^2)	Basin No.	Basin name	Total depressions area (km^2)
1	Ghouimer	6.52	16	Selsli	0.35
2	Om El-Hableen	2.13	17	Muwaieha	1.81
3	Basin # 3	0.35	18	Basin # 18	0.73
4	Daghbj	1.86	19	Basin # 19	0.53
5	El Hatiel	2.14	20	Basin # 20	1.04
6	Basin # 6	0.12	21	Abou Je'Alah	1.54
7	Basin # 7	1.05	22	Al A'ayah	0.88
8	El Assla	11.24	23	Ed-Dowikhlah	1.17
9	Basin # 9	0.83	24	El-Baghdadi	0.46
10	Mreikh	2.19	25	Ketanah	0.37
11	Kawes	2.88	26	Basin # 26	0.24
12	Osheer	1.22	27	Esh-Shoabaa	0.87
13	Basin # 13	0.24	28	Basin # 28	1.43
14	Methweb	3.05	29	Da'af	0.34
15	Ghlil	1.34			

4.3 Geometric analysis of watersheds

Geometric shaping of water basins is considered as a one of the key factors that control the mechanism of geographic distribution of water on the surface of the earth. In this study, geometric standards are considered for water basin perimeter without taking into account morphometric standards located within this basin. Thus, the shape of the water basin, as well as the elongation rate and rotation rate are the most important standards.

4.3.1 Shape

There are several patterns that characterize the outer shape of water basin, each of them play a role in run-off from the upstream to downstream. The basins shape control the branching process of tributaries and thus the flow regime (Black, 1991). For example, basinswith circular shape has regular water discharge from all available tributaries and consequently water can reach the outlet at the same time. But this is not the case in basin with oval-like shape basins where there are differences in the discharge process according to the timing and therefore is considered as less prone to floods. There are several characteristics for specifying the shape of water basins, and the most important are as follow:

4.3.1.1 Length/Width ration

The ratio between the basin length and width (LW) is an indicator of surface flow effectiveness, and consequently an increase in the ratio of the width compared to the length make longer the flow duration and vice versa. Where the average value of this ratio in regular basins is 0.5, which means that the basin length is equal to the twice of its width. This is defined by the following simplified equation:

$$\text{Length / Width ratio} = \frac{\text{L} = \text{length of the basin}}{\text{W} = \text{Width of the basin}}$$

4.3.1.2 Shape factor (S_f)

The controlling factor in shaping water basin is determined according to the central length (Lca), which is the length of the main channel (Figure 7). It largely represents the basin orientation, and has an integral role in the flow mechanism from different tributaries to the major outlet. It also contributes to infiltration rate and time duration of run-ff to be joined from different existing branches with different dimensions. Thus, the following equation is used to elaborate this factor.

$$S_f = (LL_{ca})^{0.3}$$

Where L is the maximum length of water basin Lca is the distance from the center of the basin circle to the outlet.

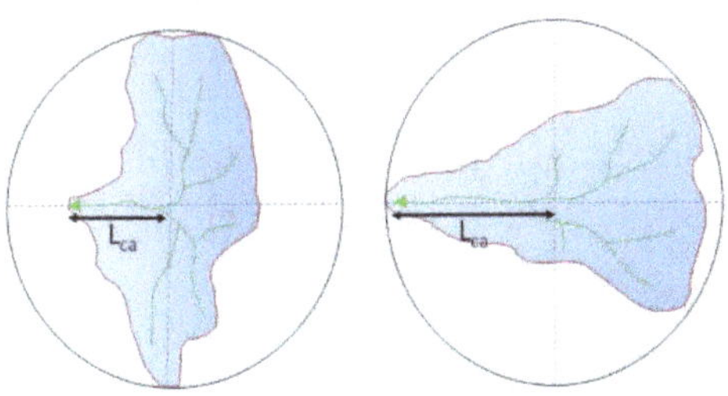

Fig. 7. Representation showing the shape factor in surface water flow regime.

4.3.1.3 Width/outlet width ration

Width/outlet width ration is the ration between the basin width to the outlet width. If the basin width from the upstream is bigger enough than the outlet width, flood occurrence is more probable (Al Saud, 2010a a, b), as shown in Figure 8. This can be attributed to the inadequate drainage capacity of water derived from upstream; and therefore made impossible the regular drainage mechanism through the outlet.

The width/outlet width ration is almost consistent with the characteristics that can be extracted from the DEM, as mentioned previously. However, in this case we treat the basin outer limits and not with the terrain features inside. It can be simply represented as:

$$\frac{W_b = \text{Basin width}}{W_d = \text{Outlet width}}$$

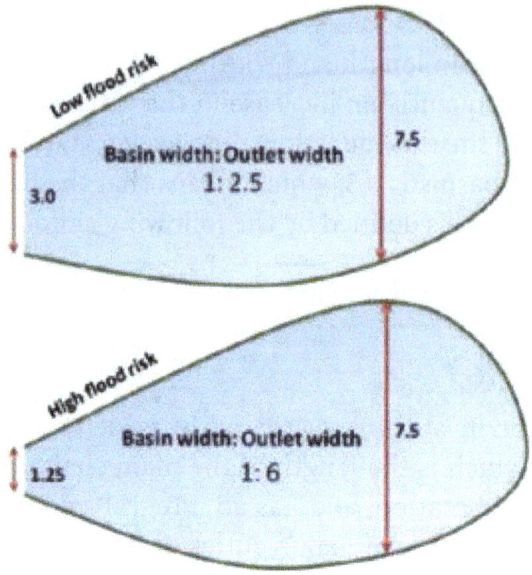

Fig. 8. Representative figure showing the widths of upstream and out let.

4.3.1.4 Elongation ration

It also essential to characterize the basin elongation, or one-direction stretching, according to its area. It equals 1 for the intake circle, whilst, it is zero for the straight line. According to Schumm (1956), the elongation ration equals:

$$R_e = 2/L_m (A/\pi)^{0.5}$$

where Lm is the basin length parallel to the primary watercourse, and A is the basin area, as shown in Figure 9. Therefore elongation ration governs the flow regime at the outlet.

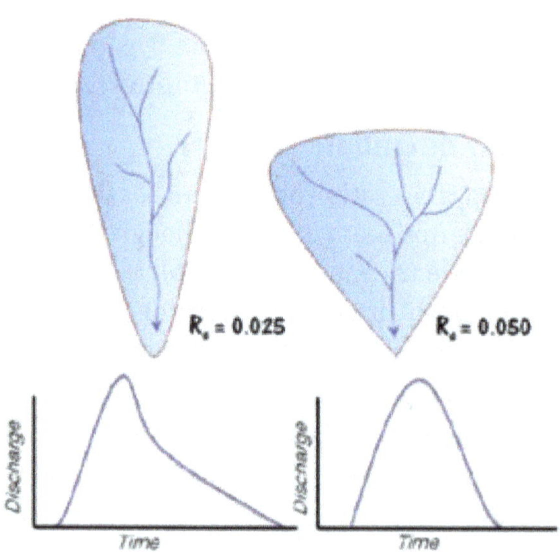

Fig. 9. Representation for two basins with different elongation ratio.

4.3.1.5 Circularity ration

Circularity ration (Rc) represents the uniformity between the external perimeter of the water basin comparing to its circular form (Miller, 1953).

$$Rc = A/A0$$

Where A0 is the perimeter of the circle, A area of the basin

The analysis of principal geometric characteristics was applied on the main water basins in the area of study with the exception to none completed basins (joining basins) as shown in Table 7. It is shown some basins are characterized by normal effectiveness with respect to this type of natural disasters.

Table 7. Shape characteristics in water basins as shown from DEM.

Basin No.	Basin name	Length/width ratio (Lw)	Shape factor (Sf)	Width/outlet width (Wb/Wd)	Elongation ration (Re)	Circularity ration (Rc)
1	Ghouimer	1:0.87	5.83	1:2.5	0.82	3.30
2	Om El- Hableen	1:2	3.88	1:1.8	0.68	1.39
4	Daghbj	1:3	4.42	1:2	0.51	1.03
5	El Hatiel	1:4.3	4.72	1:3	0.44	0.96
7	Basin # 7	1:3	3.22	1:2	0.53	0.60
8	El Assla	1:1.2	6.77	1:8.5	0.58	2.36
10	Mreikh	1:1.75	3.64	1:1.3	0.78	1.08
11	Kawes	1:3.8	5.99	1:2.57	0.38	0.85
12	Osheer	1:1.6	2.81	1:3	0.58	0.62
14	Methweb	1:2.2	5.26	1:5	0.43	0.82
15	Ghlil	1:2.15	3.45	1:2.3	0.47	0.57
17	Muwaieha	1:1.55	3.05	1:1.6	0.63	0.72
21	Abou Je'Alah	1:1.5	2.57	1:1.7	0.91	0.80
22	Al A'ayah	1:2	3.25	1:1	0.45	0.48
23	Ed- Dowikhlah	1:1.6	2.92	1:1.55	0.61	0.58
24	El- Baghdadi	1:2.2	3.34	1:1.4	0.59	0.91
25	Ketanah	1:1.7	3.78	1:2.3	0.56	0.80
27	Esh- Shoabaa	1:2.5	4.16	1:1.5	0.49	0.80
29	Da'af	1:1.75	3.58	1:1.45	0.62	0.93

4.4 Morphometric analysis of drainage systems

4.4.1 Drainage density

Water channels take several geometric patterns, and compose a collection of natural networks. However, the most important in drainage distributions is the density of tributaries among the basin. For example, we can find areas with high drainage density faced with areas with a lower density. This is primarily due factors related to the terrain characteristics. Also, drainage density can be calculated using the following equation:

$$\frac{\sum L}{A} = \frac{\text{Total length of tributaries in a particular area}}{\text{Surface of the area}}$$

Table 8 shows the resulting values of drainage density in each basin.

Table 8. Drainage density in the water basins.

Basin No.	Basin name	Density (Km/Km2)	Basin No.	Basin name	Density (Km/Km2)
1	Ghouimer	0.98	16	Selsli	0.53
2	Om El-Hableen	1.12	17	Muwaieha	1.02
3	Basin # 3	0.28	18	Basin # 18	1.13
4	Daghbj	0.96	19	Basin # 19	0.97
5	El Hatiel	1.36	20	Basin # 20	1.20
6	Basin # 6	041	21	Abou Je'Alah	1.53
7	Basin # 7	1.15	22	Al A'ayah	1.27
8	El Assla	1.77	23	Ed-Dowikhlah	1.24
9	Basin # 9	0.61	24	El-Baghdadi	1.14
10	Mreikh	1.76	25	Ketanah	1.25
11	Kawes	1.37	26	Basin # 26	1.23
12	Osheer	0.54	27	Esh-Shoabaa	2.20
13	Basin # 13	0.35	28	Basin # 28	1.15
14	Methweb	1.33	29	Da'af	0.95
15	Ghlil	0.85			

4.2.2. Stream order

Usually, the biggest fraction of the surface runoff occur in the channel and its different tributaries to form a water network starting from reaches from the highest elevations to reach at the end the outlet. From this point, the relation between different stream orders is seen. Small streams that starts from the top of the mountains and that are connected from one side are given the 1st order appellation. The 2nd stream order is created by the junction of two or more 1st -order streams, and so on.

Hence the process of identifying different stream orders is not the purpose of this study, but it is a kind of tools that help us in analyzing the relation between the number of these streams and more certainly the bifurcation ratio which is calculated from the following equation:

$$B_r = N_r/N_{r+1}$$

Where "Nr "is the number of streams for the order "r", and "N_{r+1}" is the number of streams for a higher order.

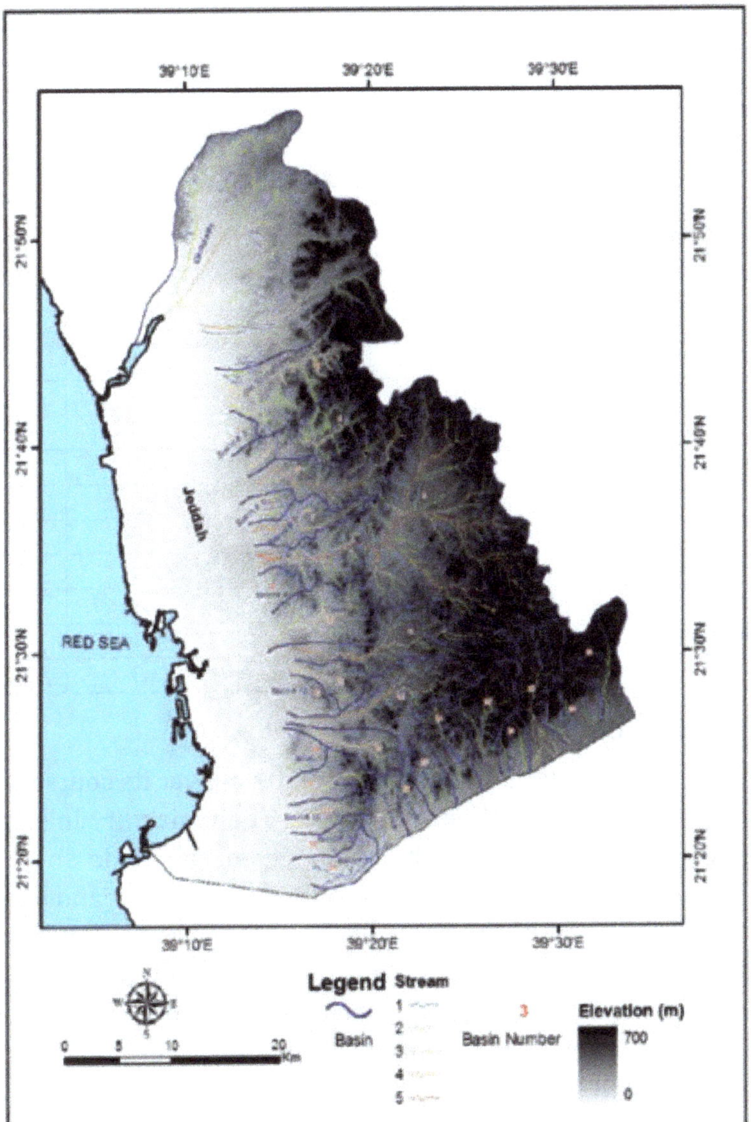

Fig. 10. Stream orders in the basins of the area of study.

In this study, the Arc-GIS 9.3 software was used to classify streams in their specific orders (Figure 10), followed by calculating the morphometric variables needed where all streams and their tributaries are numbered, making the calculations easier (Table 9).

Table 9. Major variable for stream orders in the area basins.

Basin No.	Basin name	Number of stream orders 1	2	3	4	5	Total length (km)	Bifurcation ratio (Br)
1	Ghouimer	107	68	15	8	-	342.7	2.64
2	Om El- Hableen	45	24	26	-	-	148.9	1.39
4	Daghbj	37	24	14	-	-	101.7	1.62
5	El Hatiel	73	24	16	8	-	117.5	2.18
6	Basin # 6	14	5	3	-	-	16.2	2.23
7	Basin # 7	30	10	7	3	-	51.9	2.25
8	El Assla	351	135	54	38	14	589.4	2.30
9	Basin # 9	7	7	-	-	-	12.8	1.00
10	Mreikh	31	23	14	2	-	74.9	3.33
11	Kawes	52	24	12	7	-	117.3	1.96
12	Osheer	7	7	5	-	-	20.2	1.20
14	Methweb	56	16	6	7	-	87.3	2.34
15	Ghlil	17	3	7	-	-	37.8	2.41
17	Muwaieha	17	9	13	1	-	37.9	5.19
21	Abou Je'Alah	30	11	5	-	-	45.8	2.46
22	Al A'ayah	13	6	-	-	-	25.6	2.16
23	Ed-Dowikhlah	18	10	-	-	-	29.6	1.80
24	El-Baghdadi	31	14	5	-	-	41.4	2.50
25	Ketanah	48	26	8	2	-	74.7	3.03
27	Esh-Shoabaa	55	20	25	-	-	88.1	1.77
29	Da'af	18	12	-	-	-	41.3	1.50

4.4.3 Meandering ration

A stream channel has numerous meandering features over its course, with various flow characteristics occurring at each meander. The variability in slope gradient, rock types, geological formations, and other geomorphologic and hydrological features play a significant role in determining this. The meandering ratio has been seen to have an impact on water floods. This is because an increase in the meandering ratio leads to a decrease in flow energy and an increase in stream load capacity. This is due to the erosion mechanism that occurs at the meandering locations. The meandering ratio is computed using the following equation and the results are presented in Table 10.

$$\frac{\text{Primary stream length (curved)} = L_m}{\text{Primary stream (straight)} = L_s}$$

Table 10. Meandering and intersection ratio for the basins in the study area.

Basin No.	Basin name	Meandering ratio (Mr)	Intersection Ratio (Ir) Km^2
1	Ghouimer	1.05	3.2
2	Om El-Hableen	1.35	6.3
3	Basin # 3	1.18	70
4	Daghbj	1.31	6.1
5	El Hatiel	1.02	5.5
6	Basin # 6	1.01	0.8
7	Basin # 7	1.04	1.2
8	El Assla	1.27	7.6
9	Basin # 9	1.10	2.4
10	Mreikh	1.23	2.6
11	Kawes	1.28	3.1
12	Osheer	1.19	0.53
13	Basin # 13	1.03	0.26
14	Methweb	1.17	3.4
15	Ghlil	1.14	1.9
16	Selsli	1.01	0.34
17	Muwaieha	1.12	2.4
18	Basin 18	1.12	1.61
19	Basin 19	1.05	1.15
20	Basin 20	1.12	1.04
21	Abou Je'Alah	1.19	4.6
22	Al A'ayah	1.17	2.04
23	Ed-Dowikhlah	1.20	1.92
24	El-Baghdadi	1.14	1.80
25	Ketanah	1.23	3.61
26	Basin # 26	1.12	2.03
27	Esh-Shoabaa	1.27	5.8
28	Basin # 28	1.17	1.17
29	Da'af	1.20	2.11

4.4.4 Intersection ration

When analyzing the morphometric properties of drainage systems, the analysis is sometimes complicated by the presence of multiple formulas and ideas.

Nevertheless, several features are deemed highly significant but receive less attention. One of the features is the intersection ratio, which may be determined by identifying the points where distinct channels and branches meet, such as confluences and diversions (Al Saud, 2009).

The quantity of junction nodes indicates a greater density of drainage and, thus, a more even flow. Typically, they are quantified by counting the nodes contained within a defined region. The results of the insertion ratio are displayed in Table 10.

5. CONCISE SUMMARY

Floods, which have a devastating impact, are mostly influenced by the geomorphological features of the surrounding landscape. Consequently, it is a widely used standard to evaluate the behavior of the terrain and its ability to accept and drain rainfall water. It is essential to prioritize the examination of all necessary geomorphological components, with a particular focus on the overall drainage system, which encompasses the catchment area and the features of the current streams within the catchment. However, when there is insufficient data and knowledge on the drainage system in a particular area, different techniques for analysis are often employed. Specifically, remotely sensed tools such as satellite photos and Geographic knowledge System (GIS) tools have played a significant role in this regard. This study included advanced tools and specific software for data manipulation and analysis. Additionally, a collection of satellite pictures and thematic maps were deployed to address the subject matter.

Accordingly, three major parameters were analyzed and then they were correlated to their response to water flow regime and thus floods. This was applied to Jeddah city, which has witnessed to severe flood events, and the hydrologic and geomorphologic measures are still unclear. These parameters include: the geometric analysis for digital elevation models (DEM), geometry of basins and drainage morphometry of streams. This will enable assessment the vulnerability of the basins in the study area to floods, as well as these parameter can be well utilized in applying flood control management approaches.

According to the assessed geometric measurements, there are several basins that have relatively big values for certain factors, such as Ghouimer in terms of shape factor, elongation, and circularity ratio. The Al-Assla basin exhibits a clear and pronounced response to flooding, mostly due to its high width-to-length ratio of 8.5:1. Nevertheless, there are many basins that have a lower level of responsiveness to floods, like the Abou Je'Alah basin.

The morphometry of streams in these basins plays a crucial role in regulating the flooding process. This phenomenon is most evident in locations where streams are linked, especially those streams that have extensive layers of silt. Such streams are found in multiple valleys within the research area. Furthermore, the specific area of interest was observed to have highly concentrated drainage systems, unlike other coastal regions in western Saudi Arabia. This indicates a distinct pattern of surface water flow through interconnected streams.

Furthermore, the growth of urban communities in the valleys disrupts most of the criteria that were investigated. Consequently, human settlements are causing the closure of many basins at their outflows. In addition to these settlements, the extensive excavation procedures and engineering particles have caused the release of sediments. These sediments are easily transported by water, leading to an increase in the amount of sediment carried by the flooded water.

REFERENCES

Al-Saud, M. (2009). Morphometric Analysis of Wadi Aurnah Drainage System, Western Arabian Peninsula, *The Open Hydrology Journal*, Vol.3, pp. 1-10

Al Saud, M. (2010a). Applying Geoinformation Techniques in Studying Flood and torrents in Jeddah Region, 2009 *(In Arabic). Arab Journal for Geoinformation*. 3, 1, 2010.

Al-Saud, M. (2010b). Assessment of Flood Hazard of Jeddah Area 2009, Saudi Arabia. *Journal of Water Resource and Protection (JWARP), Vol.* 2, No. 9, pp. 839-847

Black, P. (1991). Watershed Hydrology. *Prentice Hall Advanced Reference Series, NJ*, pp. 324 EM-DAT: The OFDA/CRED. (2010). International Disaster Data base. *Université catholique de Louvain, Brussels, Bel.* Data version: v11.08. Available from www.emdat.be/, 2010.

Es-Saeed, M., Sen, Z., Basamad, A., Dahlawi, & A. Al-Bardi, W. 2004. Strategic groundwater storage in Wadi Naáman, Makka region, Saudi Arabia. *Technical Report (in Arabic), Saudi Geological Survey-TR-2004-1*, pp. 32p.

IPCC Report. (March 2007). *The Fourth Assessment Report (AR4), March 14th 2008*, Available from : http://www.ipcc.ch/

Italconsult. (1967). Water supply survey for Yeddah-Makkah-Taif area. *Special Report. Geological investigation- Ministry of Agriculture and Water*, No.3.

KACT (King Abdulaziz City for Science and Technology). (2011). Production of digital elevation model (DEM/DTM) with high resolution using stereoscopic satellite imagery and the deduction of drainage network and its direction and the water bains for Jeddah regions. *Technical report*, pp. 1-62

Khosrowpanah, Sh., Heitz, L., Wen, Y., Park., M. (2007). Developing a GIS-based soil erosion potential model of the Ugum Watershed, *Technical Report, Water and Environmental Research Institute (WERI) of the western Pacific*. University of Guam, pp. 1-103

Miller, V. 1953. A Quantitive Geomorphic Study for Drainage Basin Characteristics in the Clinch Mountain Area, Virginia and Tennessee, *Technical Report No.3*, I-30, N6 ONR 271-30, Geology Depart., Colombia University, n.d.

Morisawa, M. (1976). Geomorphology Laboratory Manual, *John Wiley & Sons Inc. N.Y.*, pp.1- 253

Schumm, S. 1956. The elevation of drainage systems and slopes in badlands at Perth Amboy, New Jersey. *Geol. Soc. Amer. Bull.*, Vol. 67, pp. 597-646

Subyani, A., Qari, M., Matsah, M., Al-Modayan, A. & Al-Ahmadi, F. (2009). Utilizing remote sensing and GIS technologies to produce hydrological and environmental hazards in some Wadis, western Saudi Arabia (Jeddah-Yanbu). Dept of Hydrology. King Abdulaziz City for Science and Technology. General Directorate of Research Grants Program. Kingdom of Saudi Arabia.

Tribe, A. (1991). Automated Recognition of Valley Heads from Digital Elevation Models, *Earth Surface Processes Landforms*, Vol. 16, pp. 33-49.

Yehia, M. & El-Ater, J. 1997. Flood and mechanism of confronting tisk on the Egypt red sea coast. National Authority for Remote Sensing and Space Sciences. *Scientifc report submitted to the Red sea Governorate*, pp. 1-294

08

AN ANALYSIS OF TWO NONLINEAR RADIATION MODELS FOR AGRICULTURAL SUBSURFACE DRAINAGE IS CONDUCTED

1. INTRODUCTION: AN OVERVIEW

The primary purpose of subsurface agricultural drainage systems is to prevent the formation of a moisture environment that is detrimental to crop growth. This is achieved by removing shallow groundwater and promptly draining excess water caused by over-irrigation, rainfall, infiltration in channels, and underground streams. To achieve effective water evacuation, it is essential to understand the movement of water in the soil. This knowledge helps determine the amount of water that can be drained from the porous medium through drainage, which in turn affects the level of the aquifer.

An agricultural drainage system's variables can be determined by analyzing the mass and energy exchanges that take place in the soil. The analysis of these transfers, albeit more intricate due to the examination of nonlinear processes occurring in a medium with varying properties throughout time and space, can be conducted by considering the following analysis scales:

At the microscopic scale, the Poiseuille law, derived from the Navier-Stokes equations, is used to estimate the mean velocity of water flow in each soil pore. Additionally, the pressure in each pore is approximated using the Laplace equation. This analysis scale is highly suggested for a comprehensive comprehension of the fundamental mechanisms involved in water transfer activities within the soil.

At the macroscopic scale: The intricate nature described by the precise definition of the geometric structure of the pore space necessitates a shift in scale for the microscopic analysis. This shift involves introducing the concept of the representative volume element (RVE), which enables us to establish an equivalence between the actual dispersed porous medium and an imaginary continuous porous medium. In this analysis scale, the mean velocity in pores filled with water or macroscopic flow is estimated using the Darcy-Buckingham law (1907), which considers a range of pore sizes. The water pressure associated with the set is estimated using the Laplace law, applied to a larger pore size. The transfer equation that corresponds to this is commonly referred to as Richards equation (1931).

Megascopic Scale: This analysis scale is used to study a specific set of soils. It involves estimating the average velocity of water flow, known as megascopic flow, using the Darcy law. The flow is then averaged based on the Dupuit-Forcheimer hypothesis, which considers a hydrostatic pressure distribution. Additionally, water pressure is measured using piezometers. The Boussinesq equation of agricultural drainage is the appropriate transfer equation.

The analysis of agricultural drainage is typically conducted using either the Richards equation (Zaradny & Feddes, 1979; Fipps & Skaags, 1986; Saucedo et al., 2002) or the Boussinesq equation (Dumm, 1954; Pandey *et al.,* 1992; Gupta et al., 1994; Samani *et al.,* 2007). The Richards equation enables the description of transfer processes in both the saturated and unsaturated zones of the soil. However, its practical application is limited when it comes to larger scales such as an irrigation district or a farm field. This limitation arises from the challenges and expenses associated with conducting experimental work to accurately depict the soil's hydrodynamic characteristics, specifically the moisture retention curve and the hydraulic conductivity curve. Additionally, the calculation of three-dimensional water movement in the soil requires significant effort. The limitations mentioned have resulted in the predominant use of the Boussinesq equation for analyzing agricultural drainage. This approach simplifies the consideration of transfers in the soil's unsaturated zone and requires less data compared to the Richards equation. However, it still adequately describes the dynamics of water in the saturated layer of the soil.

Two recent mechanistic models have been created to enhance the conventional assumptions of agricultural drainage models found in literature. Zavala et al. (2005) have created a model for agricultural subsurface drainage using the two-dimensional Richards equation. The above differential equation is subject to a nonlinear radiation condition at the boundary of the drains. This representation accurately captures the mass and energy exchanges in a drainage system. However, Zavala *et al.* (2004) and Fuentes et al. (2009) have examined agricultural drainage using the Boussinesq equation. They have respectively determined the appropriate boundary condition to be applied in agricultural drains using this equation, and established the connection between the moisture retention curve and the storage coefficient in shallow unconfined aquifers.

The objective of this chapter is to introduce and analyze the numerical solutions of the two previously mentioned models. Specifically, we will compare the mass and energy transfers obtained using the Richards equation and the Boussinesq equation. It is important to note that both equations are susceptible to nonlinear radiation conditions in the drains.

2. MATERIALS AND METHODS

2.1 Macroscopic scale

Soil water movement in a subsurface drainage system is a three-dimensional phenomenon, for which reason its description should be made using the Richards equation in three dimensions; nonetheless, due to the effort of calculation that the resolution of this form of Richards equation entails, it is convenient to accept the hypothesis that the phenomenon is basically two-dimensional (that is to say, it is made according to planes perpendicular to the direction of the drain). If, in addition, it is assumed that the water uptake by plant roots is negligible, the two-dimensional form of the Richards equation may be written as follows:

$$C(\psi)\frac{\partial \psi}{\partial t} = \frac{\partial}{\partial x}\left[K(\psi)\frac{\partial \psi}{\partial x}\right] + \frac{\partial}{\partial z}\left[K(\psi)\left(\frac{\partial \psi}{\partial z} - 1\right)\right] \quad (1)$$

where ψ is the pressure head $[L]$; $C(\psi) = d\psi(y)$; dy is the specific water capacity $\left[L^{-1}\right]$; $q(\psi)$ is the volumetric water content $\left[L^{3}L^{-3}\right]$ $K(\psi)$ is the hydraulic conductivity $\left[LT^{-1}\right]$ x and z are, respectively, the horizontal and vertical coordinates $[L]$; and t is time$[T]$.

If a drainage system with equidistant parallel pipes installed at the same depth is considered, it is possible to define a domain for equation (1) as with the one shown in Fig. 1.

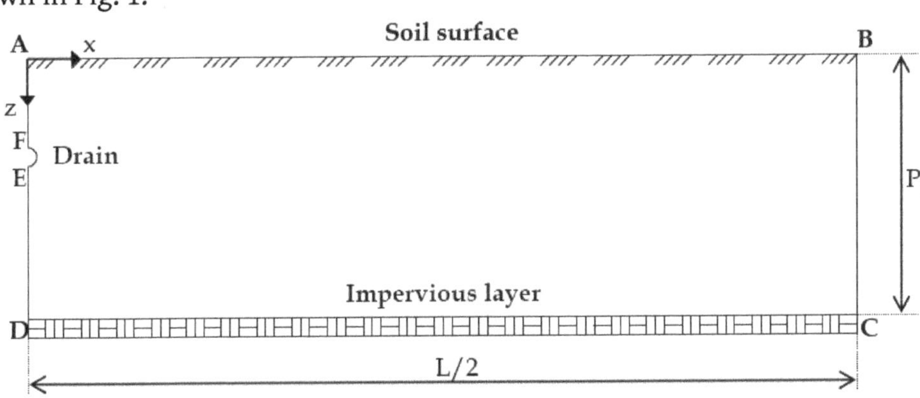

Fig. 1. Domain for the Richards equation.

The description of agricultural subsurface drainage by equation (1) requires the definition of the initial status of the pressure head in the porous media as well as its boundary conditions. The initial condition of the water pressure in the soil is specified as a known space function:

$$\psi = \psi_i(x,z) \tag{2}$$

By the flow symmetry, it is known that the Darcy flow in a perpendicular direction to segments AF, DE and BC is null (Neumann boundary condition), and a similar situation occurs in the boundary segment CD due to the impervious layer.

$$-K(\psi)\frac{\partial \psi}{\partial x} = 0 \quad x = 0; \quad z \in \overline{AF} \text{ y } z \in \overline{DE}; \quad t > 0 \tag{3}$$

$$-K(\psi)\frac{\partial \psi}{\partial x} = 0 \quad x = L/2; \quad z \in \overline{BC}; \quad t > 0 \tag{4}$$

$$-K(\psi)\frac{\partial(\psi - z)}{\partial z} = 0 \quad x \in \overline{CD}; \quad z = P_E; \quad t > 0 \tag{5}$$

where P_E is the depth of the impervious layer, measured as from the soil surface; and L is the space between drains. In the soil surface (segment AB) when the rain or evaporation intensity (i) is known, a Neumann type boundary condition can be implemented:

$$-K(\psi)\frac{\partial(\psi - z)}{\partial z} = i \quad x \in \overline{AB}, \quad z = 0; \quad t > 0 \tag{6}$$

Consistent with Zavala *et al.* (2005), the soil water transfer to the drain shall be described with the following nonlinear radiation condition:

$$-K(\psi)\frac{\partial(\psi - z)}{\partial n} = q_o\left(1 - \frac{\psi}{P}\right)^{\alpha}\left(\frac{\psi - h_t}{P}\right)^{\beta} \quad \forall x, z \in \overline{EF}; \quad t > 0 \tag{7}$$

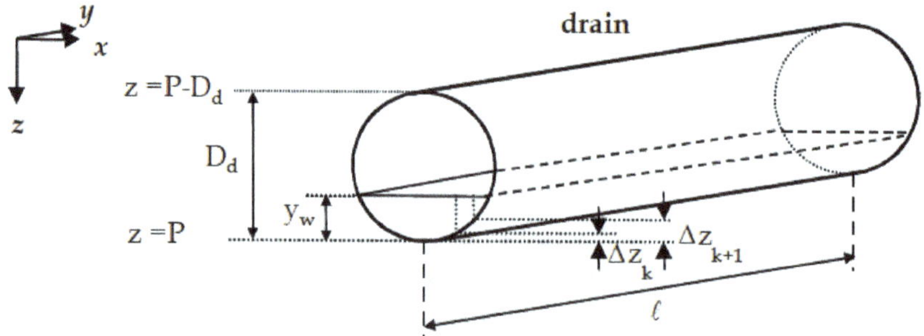

$$h_t = \begin{cases} 0 & P - D_d \leq z < P - y_w \\ y_w - \Delta z_i & z \geq P - y_w \end{cases} \qquad y_w = \text{water depth}$$

Fig. 2. Scheme of the water flow in a drain.

where $\partial / \partial n$ is the normal derivative; qo is a particular value of the water flow in the soil [LT^{-1}]; σ and β are dimensionless shape parameters; P is the depth of the drain [L] ; and ht is the pressure inside the drain [L], equal to the atmospheric pressure (h_t=0) in the segment of the drain's internal perimeter in contact with the air and equal to the water depth had at every point of the drain's internal perimeter in contact with water (Fig. 2).

The application of relation (7) requires knowledge of the evolution of mean depth of the water in the drain. If the variations in the longitudinal direction are negligible, the depth of the water in the drain may be supposed to be uniform in space but variable in time. In this situation, the evolution in time of the depth of water may be calculated as from an equation that relates the flow velocity with the energy loss in the movement direction. Because of its generality, the fractal resistance law proposed by Fuentes *et al.* (2004) is used in this work:

$$V = \kappa \frac{g^d}{\nu^{2d-1}} R_H^{3d-1} J^d \tag{8}$$

where V is the water's mean velocity in the drain $\left[LT^{-1}\right]$; d is a dimensionless parameter that varies between 1 2 £ d £ 1 in terms of the type of flow (turbulent or laminar); k is a dimensionless coefficient; g is the gravitational acceleration $\left[L\,T^{-2}\right]$; n is the water kinematic viscosity $\left[L^2T^{-1}\right]$; R_H is the hydraulic radius [L] ; and J is the friction slope $\left[LL^{-1}\right]$.

The combination of relation (8) and the continuity equation for steady flow - which indicates that the flow Q is the product of the hydraulic area and the mean velocity (Q = VA) - allows the obtaining of the relation between the mean depth of the water in the drain and the water flow that leads to $Q = \kappa g^d A R_H^{3d-1} J^d / \nu^{2d-1}$. However, the application of this relation displays a limitation: there are two unknown variables (flow and water depth) and only one equation. This problem is resolved by outlining a second equation that is obtained when the nonlinear radiation condition is integrated with (7):

$$Q = 2\ell \int_\Omega q_o \left(1 - \frac{\psi}{P}\right)^\alpha \left(\frac{\psi - h_t}{P}\right)^\beta d\Omega \tag{9}$$

where Ω is the perimeter of the drain semi-circumference and ℓ is its length.

In order to model agricultural drainage with the system of equations (1-9), it is crucial to have the analytical representations of the soil hydrodynamic characteristics $q(\psi)$ and $K(\theta)$. In field and in laboratory applications, Fuentes *et al.* (1992) recommend using the van Genuchten model for the moisture retention curve (van Genuchten, 1980), subject to the Burdine restriction (Burdine, 1953):

$$\frac{\theta(\psi)-\theta_r}{\theta_s-\theta_r}=\left[1+\left(\frac{\psi}{\psi_d}\right)^n\right]^{-m} \quad m=1-\frac{2}{n} \text{ with } n>2 \tag{10}$$

where θ_s is the saturated volumetric water content; θ_r is the residual volumetric water content; $\psi_d < 0$ is a pressure scale parameter; m and n are the shape parameters.

As for the hydraulic conductivity curve, they suggest using the Brooks & Corey model (1964):

$$K(\theta)=K_s\left(\frac{\theta-\theta_r}{\theta_s-\theta_r}\right)^{\eta} \tag{11}$$

where K_s is the saturated hydraulic conductivity $\left[LT^{-1}\right]$;and η is a positive dimensionless shape parameter.

2.2 Megascopic scale

Rough descriptions of the mass and energy transfers of subsurface agricultural drainage systems can be obtained with the Boussinesq equation for unconfined aquifers. As per the hypothesis that variations in the direction of the drain are negligible and that the null recharge - the dynamic of the water in the saturated thickness of the soil - can be described with the Boussinesq equation of agricultural drainage:

$$\mu(H)\frac{\partial H}{\partial t}=\frac{\partial}{\partial x}\left[K_s(H-H_i)\frac{\partial H}{\partial x}\right] \tag{12}$$

where H and H_i are, respectively, elevations of the free surface or hydraulic head and of the impervious layer, measured from the same reference level [L] , when the impervious layer is approximately horizontal it may be supposed as the marker level and take H_i=0; $\mu(H)$ is the storage coefficient $\left[L^3L^{-3}\right]$, which, in a shallow unconfined aquifer is a function of the hydraulic head (Hilberts *et al.*, 2005; Fuentes *et al.*, 2009).

Taking into account the van Genuchten model for the moisture retention curve, subject to the Burdine restriction and the hydrostatic pressure distribution hypothesis, this allows the obtaining of the following analytical representation for the storage coefficient (Fuentes *et al.*, 2009):

$$\mu(H)=(\theta_s-\theta_r)\left(1-\left\{1+\left[(H-H_s)/\psi_d\right]^n\right\}^{-m}\right) \quad \text{with } m=1-2/n \tag{13}$$

where H_s is the soil surface elevation.

To resolve equation (12) on the domain shown in Fig. 3, it is necessary to define the initial conditions and the boundary conditions. The specification of these limit conditions is more convenient if the free surface position is counted as from the impervious layer:

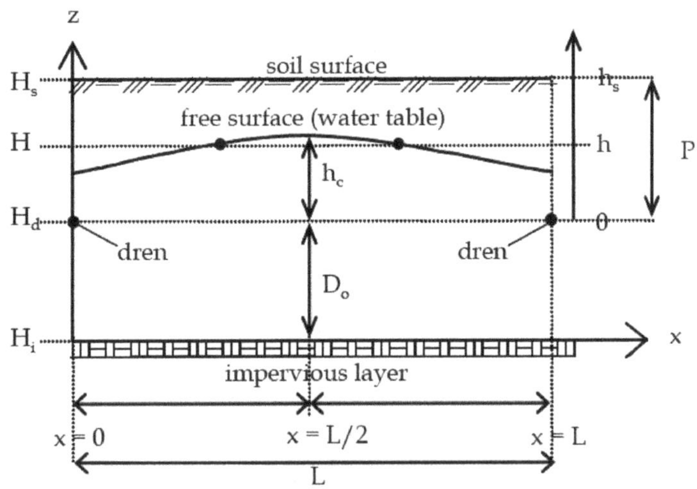

Fig. 3. Domain to the Boussinesq equation.

$$H(x,t) = D_o + h(x,t) \tag{14}$$

Where h(x, t) is the hydraulic head counted as from the position of the drains; and D_o is the depth of the impervious layer measured as from the drain's position (see Fig. 3).

In general terms, the pressure's initial condition shall be specified as the elevation of the free surface throughout the horizontal coordinate x:

$$h(x,0) = h(x) \tag{15}$$

Zavala *et al.* (2004) have shown that the relation that will subject the Boussinesq equation in the boundary of the drains is the following fractal radiation condition:

$$-K_s \frac{\partial h}{\partial x} \pm \gamma \overline{K}_s \left(\frac{h}{P}\right)^{2s-1} \left(\frac{h}{L}\right) = 0 \tag{16}$$

where γ is a dimensionless conductance coefficient; s is the quotient of the soil fractal dimension D_f and the dimensional Euclidean space ($s = D_f/3$); $\overline{K}_s$ is the hydraulic conductivity of the soil-drain interface. The positive sign in equation (16) is taken for the drain located in coordinate x = 0 and the negative sign for x = L . As per Zavala *et al.* (2007), it is convenient to express equation (16) as follows:

$$-K_s \frac{\partial h}{\partial x} \pm q_s \left(\frac{h}{P}\right)^{2s} = 0 \tag{16.1}$$

where $q_s = \gamma \overline{K}_s P/L$ ..

The quotient dimension s is implicitly defined in terms of the total volumetric porosity (ϕ) as:

$$(1-\phi)^{s}+\phi^{2s}=1 \qquad (17)$$

If μ_a represents the total areal porosity, and considering that $\mu_a=\phi^{2s}$, the equation that defines the relationship between μ_a and ma is:

$$(1-\mu_a)^{1/s}+\mu_a^{1/2s}=1 \qquad (18)$$

3. APPLICATION

The mass and energy fluxes provided by the systems of equations (1-11) and (12-18) are compared using the drainage experimental information of Zavala et al. (2004). The experiment was conducted in a drainage module constructed from acrylic sheets, containing two PVC drains (Fig. 4). The drain length are ℓ is 0.30 meters and the drain diameter is 0.05 meters. There are a total of 233 circular apertures in the drain-wall, with each opening having a diameter of 1.58 millimeters. The slope of the drain is 0.001. The drainage module has additional characteristics: L = 1.0 m, P = 1.20 m, and D_o = 0.25 m. The module was filled with a modified sample of sandy soil from the Tezoyuca region in Morelos, Mexico. The soil was sieved through a 2 mm sieve and placed in layers that were 0.20 m deep, with the aim of keeping a consistent bulk density. The soil was thoroughly saturated by applying a continuous water pressure on its surface until almost all the trapped air was eliminated. After closing the drains, the water level on the soil surface was lowered. To prevent evaporation, a plastic cover was placed over the module's surface. Subsequently, the drains were unsealed in order to quantify the volume of water that was drained over a period of ten days. It is important to note that from the beginning, the water pressure was evenly distributed and there was no replenishment of water during the drainage phase.

The hydrodynamic characterisation of the soil, executed independently of the transient drainage test, allowed the determination of the parameters of the van Genuchten model and the Brooks & Corey model (Zavala *et al.*, 2004): $\theta_s=\phi$ = q 0.539 cm^3/cm^3, θ=0 cm^3/m^3, m=0.373, η=3.767, K_s= 0.183 m/h and ψ_d=-0.418 m . To estimate the dimensionless coefficient of the fractal resistance law (equation 8), Zavala *et al.* (2005) compare this law with the Hazen-Williams relation, considering κ = 0.54 PVC drains).

The experimental conditions allow the identification of all the domain boundary segments, except the one associated with the drain, as impermeable (Neumann boundary condition). When the porosity ϕ=0.539 cm^3/cm^3 is introduced in equation (17) and this function is resolved, the exponent of the fractal radiation condition (equation 16.1) for the Boussines equation is obtained (s=0.636). Eventually, the parameters α , β and q_o intervening in the nonlinear radiation condition (equation 7), as well as scale parameter qs (equation 16.1), are determined as from the evolution in time of the experimental drained depth. To save the problem of estimating three parameters from one test only, Zavala *et al.* (2005) assume =a b and obtain $\alpha = \beta$ 1.88 and q_o = 300Ks ; in the case of the Boussinesq equation, Zavala *et al.* (2004) report $q_s = 0.913K_s$.

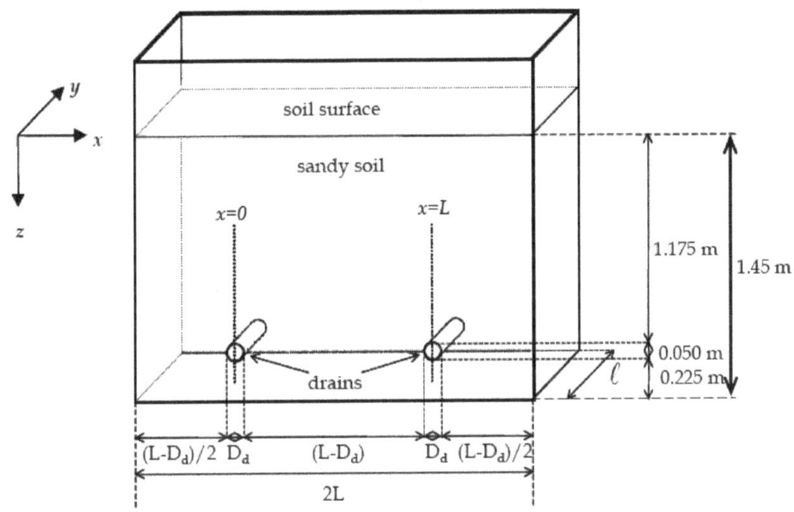

Fig. 4. Scheme of the drainage module.

To compare the transfers described, with two flow models, it is necessary to resolve the systems (1-11) and (12-18). Both systems of equations are numerically solved following the process employed by Zavala *et al.* (2004) & (2005). The spatial discretisation is carried out by using the Galerkin finite-element method; temporal discretisation is performed with a finite- difference implicit method. The resulting system becomes lineal using the Picard iterative method; the algebraic equation system is solved using a preconditioned conjugated gradient method. These methods are well documented, for example in Zienkiewicz *et al.* (2005)

The solution domain discretisation of the two-dimensional Richards equation is carried out by applying the Argus-One program, with which a finite element mesh of 10,795 nodes was generated and distributed in 21,082 elements (Fig. 5) - this being the minimum spaces Dx_{min} D $=z_{min}$ = 0.2 cm and the maximum ones Dx_{min} D $=z_{min}$ = 2.0 cm -. The solution domain of the one-dimension Boussinesq equation was discretised generating a mesh of 201 nodes and 200 finite elements of a uniform size (Δx = 0.5 cm).

By applying the numerical solutions of the systems (1-11) and (12-18), the drainage experiment is simulated to determine and to compare the evolution in time of the water depth evacuated by the drain and the corresponding variation of the free surface to half of the space between the drains. The results obtained for the mass transfer are presented in Fig. 6a and 6b and, for the energy transfer, the results are presented in Fig. 7a and 7b; in each case, as the drainage time increases, the calculated evolutions trend to a limit value because the recharge is null.

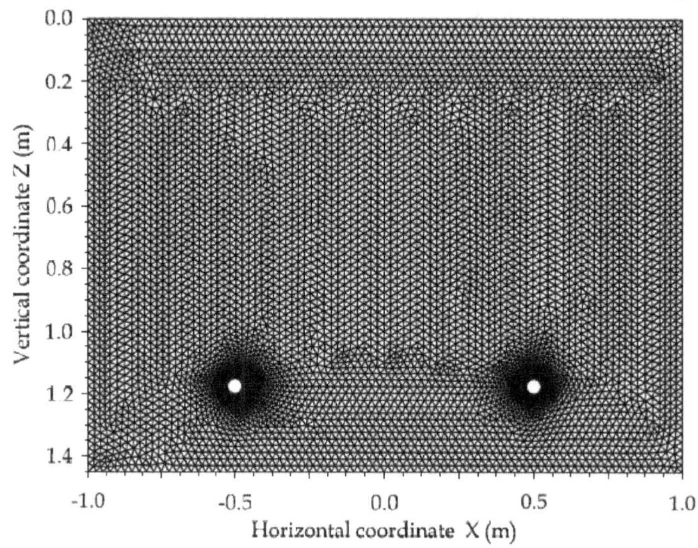

Fig. 5. Finite element mesh (Richards equation domain).

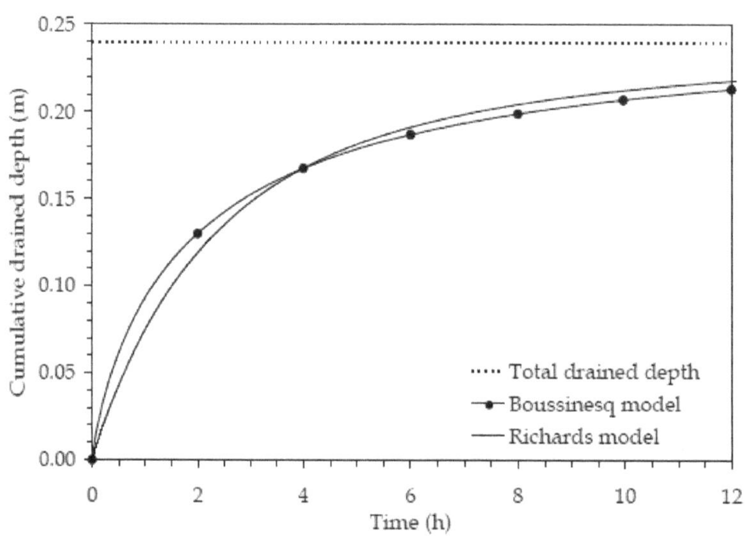

Fig. 6a. Comparison of the drained depth evolutions calculated with the Richards model (1-11) and the Boussinesq model (12-18). Twelve hours of the drainage experiment.

A good agreement between the evolution of the drained depth described with the Richards equation and the evolution obtained with the Boussinesq equation (R^2 = 0.9847) can be appreciated in Fig. 6a and 6b; this is a logical result because both evolutions are the direct product of the estimation of the radiation conditions' parameters (7) and (16.1), as from the experimental data of the drained depth. However, the evolution of the water table was not considered to optimise parameters, for which reason this variable is a good comparison element. When observing the results shown in Fig. 7a and 7b, an important discrepancy can be appreciated between the evolution described by the model based on the Richards equation as to the evolution obtained with the Boussinesq model; during all the drainage time, the water table drawdown calculated with the Boussinesq model is slower than the one obtained with the Richards model.

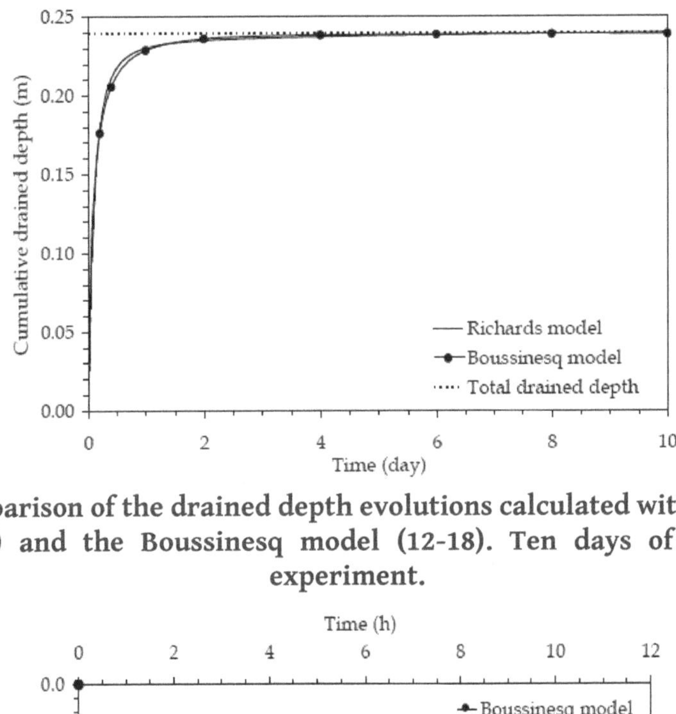

Fig. 6b. Comparison of the drained depth evolutions calculated with the Richards model (1-11) and the Boussinesq model (12-18). Ten days of the drainage experiment.

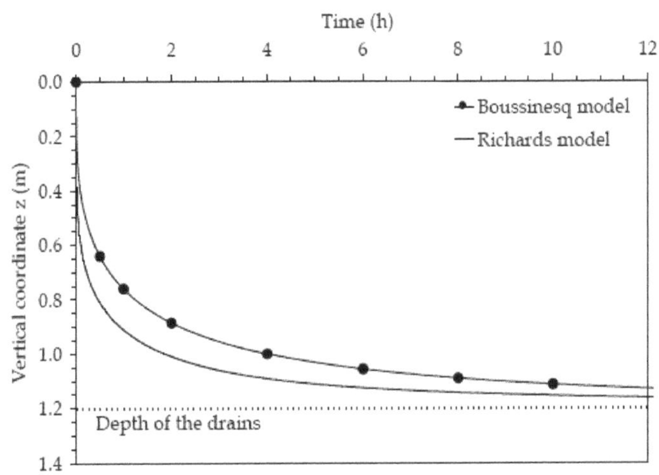

Fig. 7a. Comparison of the evolutions of the free surface to half of the space between the drains calculated with the Richards model (1-11) and the Boussinesq model (12-18). Twelve hours of the drainage experiment.

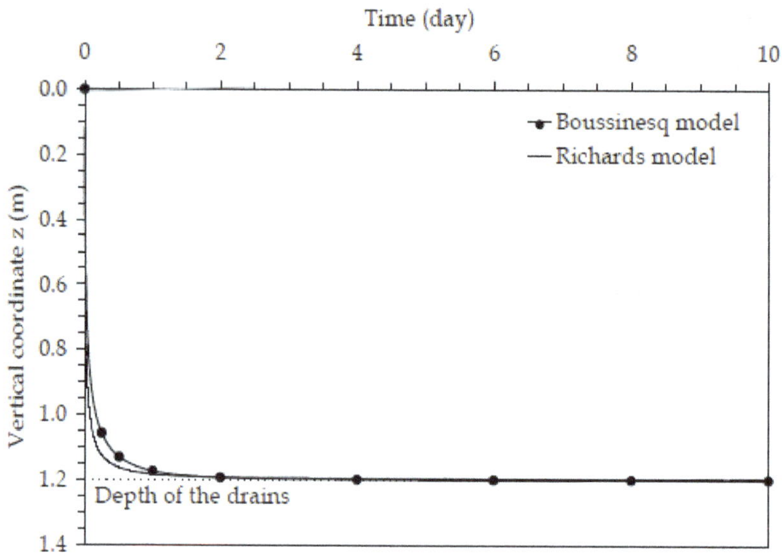

Fig. 7b. Comparison of the evolutions of the free surface to half of the space between drains calculated with the Richards model (1-11) and the Boussinesq model (12-18). Ten days of the drainage experiment.

Taking into account that the Boussinesq equation may be obtained by integrating the Richards equation in the vertical with the hypothesis of a hydrostatic pressure distribution (Bear, 1972) - circumstances that are not met in the regions closer to the drain - we have the most accurate description of the hydraulic variables in a subsurface drainage system corresponding to the one provided by the Richards equation. Considering, in addition, the simulation results, it can be seen that the model based on the Boussinesq equation (equations 12-18) cannot simultaneously reproduce the evolutions of the mass and energy transfers that are described by the model based on the Richards equation; that is to say, if it reproduces the evolution of the mass, it is not feasible that it reproduces the energy evolution or, if it reproduces the energy evolution, it cannot reproduce the mass evolution.

The limitations of an accurate simultaneous description of mass and energy transfers with the Boussinesq equation should be had in mind when it is used to estimate soil's hydraulic properties or when it is applied to the design of drainage systems. On one hand, it is traditional to consider this equation in estimating the saturated hydraulic conductivity as with the lowering or recovery measurements of the groundwater; if this were the case, the determined value would be higher than the real value of the hydraulic conductivity of the porous medium, because - as per the results of this work - the Boussinesq equation describes a minor lowering of the free surface than the one occurring in the drainage system. On the other hand, if the saturated hydraulic conductivity has been estimated according to field or laboratory tests that consider relations that are more accurate than the Boussinesq equation, and this value is used together with the Boussinesq equation to calculate the space between drains, it is possible to obtain separations shorter than that which is really needed to satisfy the water-table drawdown.

To illustrate both situations, the drainage results obtained with the Richards model are regarded as benchmarks, and numerical simulations with the Boussinesq model are carried out. The first case involves determining the saturated hydraulic conductivity value which allows for approaching with the Boussinesq model, with the water-table drawdown to half of space between drains as described with the Richards equation. In the second case, the saturated hydraulic conductivity value determined in the laboratory by Zavala *et al.* (2004) is taken up again, and the space between drains required by the Boussinesq model is calculated in order to approximate the water-table drawdown obtained with the Richards model.

The results obtained for the first case are shown in Fig. 8a and 8b. The first Figure shows the better approach of the Boussinesq equation compared to the Richards model, as to the evolution of the free surface to half of the space between drains (R2 = 0.9761), obtained with a saturated hydraulic conductivity value for the Boussinesq model of Ks = 0.500 m/h ; this value is 2.78 times higher than the value determined in the laboratory by Zavala *et al.* (2004). With this, it is shown that the hypotheses considered in the derivation of the Boussinesq equation noticeably affect their Ks estimation capacity. The overestimation of the Ks value necessarily results in an overestimation of the evolution of the drained depth, as shown in Fig. 8b.

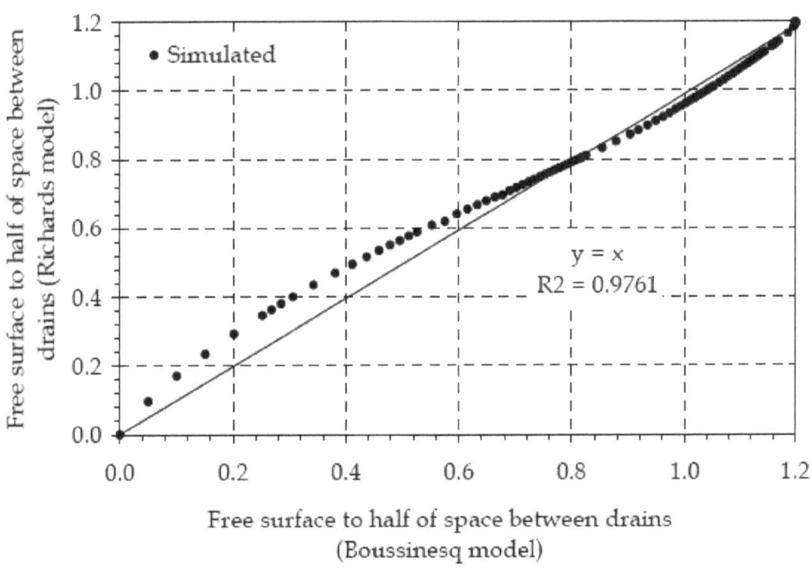

Fig. 8a. Comparison of the free surface evolution obtained by the application the Richards model with Ks = 0.183 m h , and the evolution obtained by application the Boussinesq model with Ks = 0.500 m h (best fit).

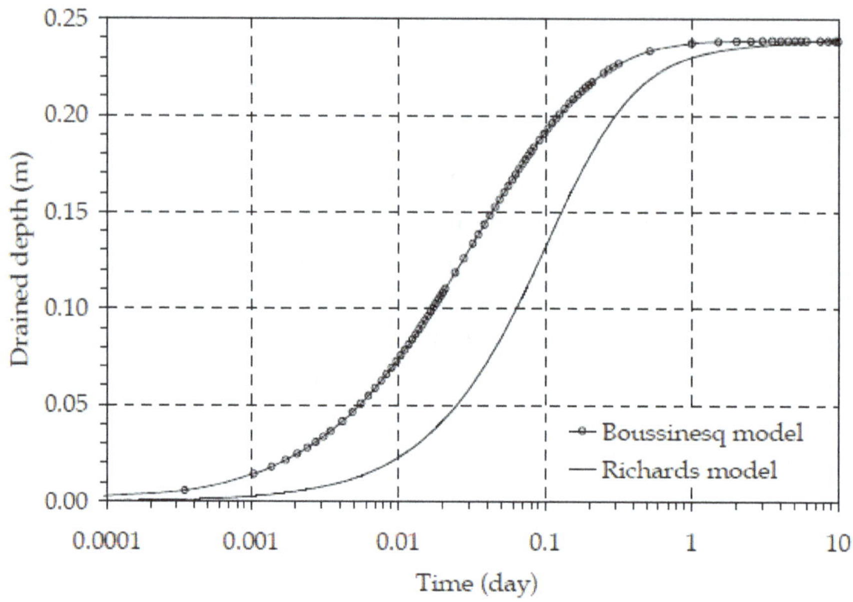

Fig. 8b. Comparison of the drained depth evolution obtained by the application the Richards model with Ks = 0.183 m h , and the evolution obtained by the application the Boussinesq model with Ks = 0.500 m h .

For case of the space between drains, it was calculated that the best approach when considering the water-table drawdown between Boussinesq model (equations 12-18) and the Richards model (equations 1-11) is obtained when the space between drains for the first model is 0.41 m (R2 = 0.9899). The difference between the real separation of the experimental drainage system considered by the Richards model and the theoretical separation required by the Boussinesq model to reproduce the lowering settled is 143%.

4. CONCISE SUMMARY

Two mechanistic models have been examined for simulating mass and energy transfers in agricultural subsurface drainage systems. The first model solves the Richards equation on a two-dimensional domain, incorporating a nonlinear radiation boundary condition along the perimeter of the drain. The second model utilizes the Boussinesq equation on a one-dimensional domain, incorporating a variable storage coefficient and a fractal radiation condition within the drain.

By utilizing experimental drainage data from an unconfined aquifer with no vertical recharge, the effectiveness of both simulation models has been assessed. It has been determined that the Boussinesq equation is unable to accurately replicate both the mass and energy transfers that the Richards equation can provide. If the Boussinesq equation is employed to replicate the mass evolution represented by the Richards equation, it is certain that the Boussinesq equation will depict an energy evolution that is slower than the one presented by Richards. However, when attempting to replicate the energy changes represented by the Richards equation using the Boussinesq equation, it results in an overestimation of the mass changes related with the Richards equation.

Studies have demonstrated that the Boussinesq equation's description limits lead to an overestimation of the saturated hydraulic conductivity when using this equation to characterize soil hydrodynamics. Similarly, if the saturated hydraulic conductivity is estimated using the Richards equation, it can result in an overestimation of the spacing between drains.

The issue with simultaneously describing mass and energy transfers using the Boussinesq equation arises from the assumption of a hydrostatic pressure distribution made during its derivation. However, this assumption is not valid near the drains, as the flow lines in this area exhibit significant curvature.

It is recommended to simultaneously determine the parameters of a drainage system and estimate the hydraulic properties of the porous media using the Boussinesq equation in the study of agricultural drainage. This should be done considering the optimization procedure, the evolution of drained depth, and variations in the water table. By doing so, the effect of the hypothesis used in deriving the Boussinesq equation can be evenly distributed among the adjustment parameters. This will result in a more accurate description of mass and energy transfers in a subsurface drainage system.

The numerical models were utilized by taking into account a single laboratory drainage test. Both models can be equally used to describe drainage systems built in real agricultural environments. The only requirement is to accurately characterize the soil's hydrodynamics, which can be done using either direct methods or indirect methods (inverse issues). If the hydrodynamic properties of the soil, such as the moisture retention curve and hydraulic conductivity curve, are accurately determined in the field and the models are used as outlined in this study, the outcomes will closely resemble those provided in this work.

If the characterization of soil hydrodynamics is conducted using an inverse method, it is advisable to utilize the procedure developed by Fuentes, as outlined in Saucedo et al. (2002) and Zavala et al. (2003). This procedure is based on the volumetric porosity of the soil, the granulometric curve, and the drainage test, which can be either local or global. This methodology incorporates the principles of the Laplace law, Stoke's law, and the principles of fractal geometry. The approach is highly accurate in representing both laboratory circumstances and field conditions.

This paper presents two numerical models that adhere to the classical notion of a deterministic model. Consequently, in order to accurately describe real farm settings, it is necessary to correctly capture the spatial variability of the hydrodynamic properties of the soil.

REFERENCES

Bear, J. (1972). *Dynamics of Fluids in Porous Media.* Dover Publications, Inc., New York, 764 Brooks, R.H., & Corey, A.T. (1964). *Hydraulic properties of porous media.* Hydrol. Pap.3. Fourthv Collins, Colorado State University.

Buckingham, E. (1907). *Studies on the movement of soil moisture.* Bureau of Soils Bulletin No. 38, USDA, 1–61.

Burdine, N. T. (1953). *Relative permeability calculation from size distribution data.* Pet. Trans. AIME, 198, 71-78.

Dumm, L. D. (1954). *Drain spacing formula.* Agric. Eng. 35, 726–730.

Fipps, G. & Skaggs, R.W. (1986). *Drains as a Boundary Condition in Finite Elements.* Water Resources Res., 22, (11), 1613-1621.

Fuentes, C., Haverkamp, R. & Parlange, J.-Y. (1992). *Parameter constraints on closed-form soil- water relationships.* Journal of Hydrology, 134, 117-142.

Fuentes, C., De León, B., Saucedo, H., Parlange, J.-Y., & Antonino, A.C.D. (2004). *El sistema de ecuaciones de Saint-Venant y Richards del riego por gravedad: 1. La ley potencial de resistencia hidráulica.* Ingeniería Hidráulica en México, XIX (2), 65-75.

Fuentes, C., Zavala, M. & Saucedo, H. (2009). *Relationship between the Storage Coefficient and the Soil-Water Retention Curve in Subsurface Agricultural Drainage Systems: Water Table Drawdown.* J. Irrig. Drain. Eng., ASCE, 135 (3), 279-285.

Gupta, R. K., Bhattacharya, A.K., & Chandra, P. (1994). *Unsteady drainage with variable drainage porosity.* Journal of Irrigation and Drainage Engineering , Vol. 120 (4), 703- 715.

Hilberts, A.G.J., Troch, P.A. & Paniconi, C. (2005). *Storage-dependent drainable porosity for complex hillslopes.* Water Resour. Res., 41, W06001, doi:10.1029/2004WR003725, 1-13.

Pandey, R., Bhattacharya, A., Singh, O., & Gupta, S. (1992). *Drawdown solution with variable drainable porosity.* Journal of Irrigation and Drainage Engineering 118 (3), 382–396.

Richards, L. A. (1931). *Capillary conduction of liquids through porous mediums.* Physics 1, 318- 333.

Samani, J. M. V., Fathi, P., & Homaee, M. (2007). *Simultaneous prediction of saturated hydraulic conductivity and drainable porosity using the inverse problem technique.* J. Irrig. Drain. Eng., 133(2), 10–115.

Saucedo, H., Fuentes, C., Zavala, M. & Vauclin, M. (2002). *Una solución de elemento finito para la transferencia de agua en un sistema de drenaje agrícola subterráneo.* Ingeniería Hidráulica en México, XVII(1), 93-105.

Van Genuchten, M. Th. (1980). *A closed-form equation for predicting the hydraulic conductivity of the unsaturated soils.* Soil Sci. Soc. Amer. J., 44, 892-898.

Zaradny, H. & Feddes, R.A. (1979). *Calculation of Non-Steady Flow Towards a Drain in Saturated-Unsaturated Soil by Finite Elements.* Agricultural Water Management, 2, 37- 53.

Zavala, M., Fuentes, C. y Saucedo, H. (2003). Sobre la condición de radiación en el drenaje de una columna de suelo inicialmente saturado. Ingeniería Hidráulica en México, 18(2), 121-131.

Zavala, M., Fuentes, C. & Saucedo, H. (2004). Radiación fractal en la ecuación de Boussinesq del drenaje agrícola. Ingeniería hidráulica en México, XIX(3), 103-111.

Zavala, M., Fuentes, C. & Saucedo, H. (2005). *Radiación no lineal en la ecuación de Richards bidimensional aplicada al drenaje agrícola subterráneo.* Ingeniería hidráulica en México, XX(4), 111-119.

Zavala, M., Fuentes, C., & Saucedo, H. (2007). *Nonlinear radiation in the Boussinesq equation of agricultural drainage.* J. Hydrol., 332(3–4), 374–380.

Zienkiewicz O.C., Taylor, R.L. & Zhu, J.Z. (2005). *The finite element method. Its basis & fundamental.* Elsevier, sixth edition, 733.

09 EVALUATION OF LIME FILTER DRAINAGE SYSTEMS

1. INTRODUCTION: AN OVERVIEW

Clayey soils with a fine texture are highly fertile but typically have a low ability to transmit water (Forsburg, 2003; Tuli et al., 2005). The permeability of these soils is insufficient, resulting in frequent accumulation of water on the surface. Consequently, clay soils typically have inadequate drainage and are challenging to handle under wet conditions. The presence of sufficient drainage is crucial for agricultural activities and mechanisation to operate effectively, as stated by Ritzema in 1994. In addition, clay soils frequently exhibit a seasonal fluctuation in hydraulic conductivity due to their tendency to expand and contract (Tuller & Or, 2003). The expansion of clay hinders the penetration of rainwater into drains. It can either collect in low areas or flow out as surface runoff. Consequently, the efficiency of draining clay soils below the surface is greatly influenced by the permeability of the material used to fill the trenches. According to Ulén (2007), if we assume that 2.5% of the soil volume in a recently tile-drained clay soil consists of backfill, which is placed in a pipe trench that is 0.5 meters wide and has a drain spacing of 20 meters, it becomes clear that the method used for backfilling significantly affects the efficiency of drainage in this type of soil. Conventional underground drainage systems in these types of soils do not always guarantee the timely evacuation of surplus water. As a result, it is common practice to add amendments to the backfill of the drainage trench in order to increase the pace at which water can seep into the soil. The permeability of these soils can be significantly enhanced by filling the trenches with coarse debris. Gravel was utilized for filling drainage trenches in Western Europe, as mentioned by Ritzema (1994) and Smedema et al. (2004). Nevertheless, in certain nations, this proves to be excessively costly because to the scarcity of natural gravel supplies.

Lime additives have been found to significantly impact the physical characteristics of clay soils, as documented by several researchers (Puustinen, 2001; Rajasekaran & Narasimha, 2002; Halme et al., 2003; Mubeen, 2005; Cox et al., 2005).

They stated that by adding a specific amount of lime, the soil's bulk density can be lowered by 25%, porosity can be raised by 30%, the filtration coefficient can be enhanced by 100-300 times, the amount of aggregates (0.25-10 mm) can be increased 3 times, and the swelling can nearly disappear. Consequently, when lime is added to soil, it becomes more resistant to compaction due to the formation of hydro-silicates from calcium and magnesium carbonates. Furthermore, lime alters the properties of clay particles, causing them to clump together, which facilitates the movement of essential nutrients in plant growth, such as phosphorus, potassium, nitrogen, sulphur, calcium, and magnesium. At the same time, it reduces the presence of mobile aluminium. Thus, liming not only enhances the water regime but also alters the water quality.

Liming materials such as calcium carbonate ($CaCO_3$), gypsum ($CaSO_4{\cdot}2H_2O$), dolomitic limestone ($CaMg(CO_3)_2$), quicklime (CaO), or other calcium-containing substances can be utilized.

The typical method of applying lime involves evenly distributing it across the top layer of soil. Nevertheless, lime will not undergo a favorable reaction until it is thoroughly mixed into the soil (Doss et al., 1979). The stable structure of clay soils can guarantee the proper preparation of a soil-lime mixture. Hence, it is imperative to diligently blend the soil with liming material prior to installing lime filters and filling the drainage trenches. In order to achieve a superior soil structure, it is essential to ascertain the ideal ratio of lime. Regrettably, there is a lack of study regarding the techniques for determining the appropriate lime rates needed to enhance the infiltration characteristics of clay soils.

In modern times, the utilization of drained land must not only be economically justified, but also meet ecological standards. The evaluation of drainage's impact on water quality is inconsistent. It enhances the natural process of converting organic matter into minerals and speeds up the movement of chemical compounds from the soil to surface waters (Zucker & Brown, 1998; Busman & Sands, 2002). Hence, the primary objective of land reclamation efforts should be to minimize their detrimental impact on the environment, namely by addressing the significant release of biogenic chemicals through drainage water (Skaggs et al., 1994; Heiskanen et al., 2004).

In clay soil, the majority of runoff is discharged through drainage trenches that have higher permeability. The backfilled trench above the drain tiles can be seen as a pathway for the preferential transport of biogenic chemicals (Dekker et al., 2001; Cho et al., 2005). Preferential flow is commonly believed to have a negative effect on water quality because it transports solutes beyond the soil zone where both biological and chemical processes typically occur at their maximum rates (Ryan 1998; Ekholm et al., 1999; Van den Eertwegh, 2006). Preferential flow is typically more important in fine-textured soils compared to coarse-textured soils. Therefore, further efforts should be taken to decrease the leaching of pollutants in these soils (Van der Salm et al., 2007).

Currently, there are multiple solutions being researched to decrease the levels of biogenic chemicals in tile effluent (Foy & Dils, 1998). The introduction of calcium-containing minerals to the soil alters the ionic composition of water. Elevated soil pH affects the abundance of several dissolved ions in the soil solution. According to literature sources, the addition of lime to soil effectively inhibits the movement of phosphorus (Rhoton & Bigham, 2005). However, there is a lack of evidence regarding the changes in the concentrations of other essential ions.

In Finland, a technique has been developed for clayey soils that involves mixing burnt lime (CaO) with the backfill material used in drains (Weppling & Palko, 1994). Soils that have been treated with lime have a large concentration of soluble and exchangeable Ca2+ ions, and often include calcium carbonate CaCO3 as well. According to Tan (2010), phosphate can react with both the ionic and carbonate forms of calcium to produce molecules that are not soluble. The reactions can be depicted as follows:

$$3\,Ca^{2+} + 2PO_4^{3-} \rightarrow Ca_3(PO_4)_2\downarrow$$

$$3CaCO_3 + 2PO_4^{3-} \rightarrow Ca_3(PO_4)_2\downarrow + 3CO_2\uparrow$$

Therefore the lime filter drainage acts as a mini chemical treatment plant and may be treated as one of water pollution control methods in agriculture (Rhoton & Bigham, 2005). The result is a stable and porous backfill that efficiently binds the phosphorus in percolating water (Curtin & Syers, 2001; Murphy & Stevens, 2010). The adsorbed phosphates are held tightly and are generally resistant to leaching (Bergström et al., 2007). Phosphate binding capacity increases from sand to clay because of the adsorption of phosphorus to soil particles (Djodjic et al., 2004).

In 1995-1999, lime filter drainage (LFD) was installed on almost 1000 hectares of land in Finland as means to combat drainage-induced adverse environmental impacts of acid sulphate soils (Bärlund et al., 2005; Aström et al., 2007). The average lifetime for the LFD has been shown to exceed 10 years without any loss in treatment effect. In Sweden, the method has only been tested at one experimental site (Ulén, 2003) and the long-term effects have not been monitored.

Lithuania is located in a region with high levels of moisture, resulting in 88% of agricultural land being artificially drained. Agricultural non-point pollution is a common issue in Lithuania due to the leaching of nutrient surplus from the soil profile into the drainage water. Based on data from the Agency of the Environment Protection, it was found that in 2005, the quantities of biogenic substances exceeded the maximum allowable concentrations (MAC) in 64% of the rivers that were analyzed. According to Šileika (2010), the contribution of agriculture to the overall phosphorus burden ranges from 42.2 to 71.1%, depending on the amount of arable land. Nevertheless, significant geographical disparities and substantial fluctuations exist in the quantities of phosphorus lost from agricultural land due to disparities in soil types, soil hydrology, and land use patterns (Valsami-Jones, 2004; Withers & Haygarth, 2007). The reduction of pollutant expansion is a pressing issue in the clayey soils with limited water permeability, which constitute 45.3% of the soil coverage in Lithuania. Consequently, field tests were done to examine the effectiveness of lime filter drainage in various climate conditions and assess the influence of lime admixture in trench backfill on the quality of drainage water in clay soils.

2. STUDY OBJECTS AND INVESTIGATION METHODS

The greater part of fine-textured clay soils are located in the Middle Lithuanian Plain. Two sites at different districts were arranged for field experiments in order to assess the efficiency of lime filter drainage in such soils.

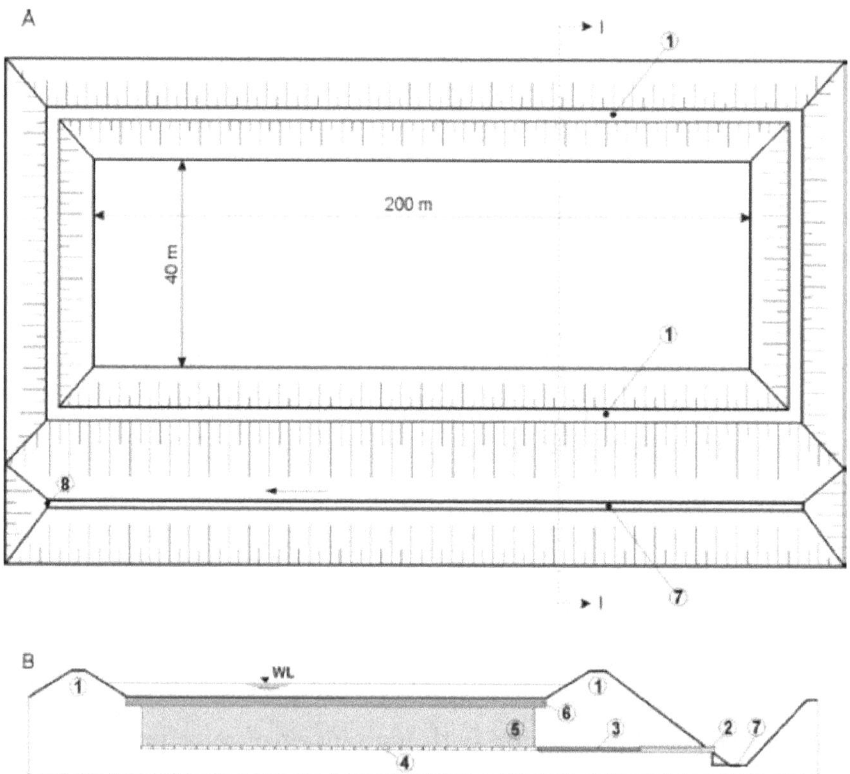

Fig. 1. Plan (A) and cross section (B) of the Pasvalys model site: 1 – dykes (h = 0.4 m), 2 – drainage outlet, 3 – non-perforated plastic pipe (l = 10 m; d = 100 mm), 4 – tile drains (d = 75 mm), 5 – trenches with different backfill, 6 – topsoil layer, 7 – open main drain, 8 – open water inlet to the pipe collector.

Pasvalys site. The site was arranged in 1986. It has been in operation for a ten-year period. The *Endocalcari-Epihypogleyic Cambisols* with soil texture of silty clay loam (according to FAO classification) are prevailing here. Soil acidity in the topsoil layer (0.0-0.2 m) ranges from pH 5.9 to pH 6.6. Deeper soil horizons were near neutrality (6.6 < pH < 7.2). The experimental set-up consisted of 200 x 40 m levelled plots surrounded by a 0.4-m high dike and resembled to an artificial land surface depression (Fig. 1). The site was drained with single tile drains, spacing at 4 m, trench width of 0.5 m, 33 cm long tile drains with an inner diameter of 75 mm were laid at the depth of 1.0 m.

Steady-state infiltration method using a ring infiltrometer was applied to determine saturated hydraulic conductivity (K-values) of the soil. The ring (H = 0.25 m, d = 0.4 m) driven into the soil at the depth of 0.1 m was supplied with a constant head of water at 0.1 m from a mariotte bottle. K-values of the topsoil layer vary within a wide range (0.65–2.75 m/d). The statistical mean value at $p<0.05$ equalled to 1.70±1.05 m/d whereas the mean value of the subsoil layer (at the depth of 0.30 m) was only 0.04±0.03 m/d.

For the investigation of the efficiency of the surface water drainage in heavy soils, drains with the higher permeability of trench backfill (when using both mineral and organic materials) were arranged. Alongside the lime filter, backfills with sand/gravel mixture (K_f=11.2 m/d), breakstone (Ø10-30 mm), turf n (decomposition rate 10%), chopped straw (5 kg/m), wood chips (0.036 m3/m) were arranged in the site. Shale ashes containing 16.8% CaO from Estonia were used as liming material. It is the rest product the burning of shale (in power stations) and contains a mixture of calcium, magnesium, potassium and other trace elements. Neutralising value of shale ashes ranges from 65 to 90% CaCO3. Four different amounts of shale ashes (0.15, 0.30, 0.40 and 0.80% of active CaO for soil mass) were mixed with the clay soil using the transporter of a multi- scoop excavator and backfilled to the drainage trenches. Control treatment was backfilled with the disturbed native soil from the trench. All treatments were installed in three replications.

During the investigations water from an adjacent stream was pumped into the experimental site. The site was flooded 16 times during the season from April to November considering soil water content.

Three sets of soil samples with 30 replications at the mid-point between the drains were taken at a depth of 0.25-0.30 m every time before the site was flooded. Soil moisture content was determined in the laboratory by the gravimetric method (the moisture content to be expressed as a percentage of the sample's dry weight). Drain discharges were measured three times a day, keeping the water level at 0.2 m above the soil surface.

The lime filter drainage effectiveness when trench backfill was mixed with different amounts of lime was assessed by comparing drainage coefficients as calculated from the following equation (Ritzema, 1994):

$$q = \frac{Q}{L\,B}, \quad (1)$$

where q – drainage coefficient, m/d, Q – pipe discharge, m3/d, L – drain spacing, m, B – length of line, m.

Kalnujai site. Another experimental site with a total area of 14.6 ha was arranged in the Jūra river catchment of the South-Western Lithuania (Fig. 2).

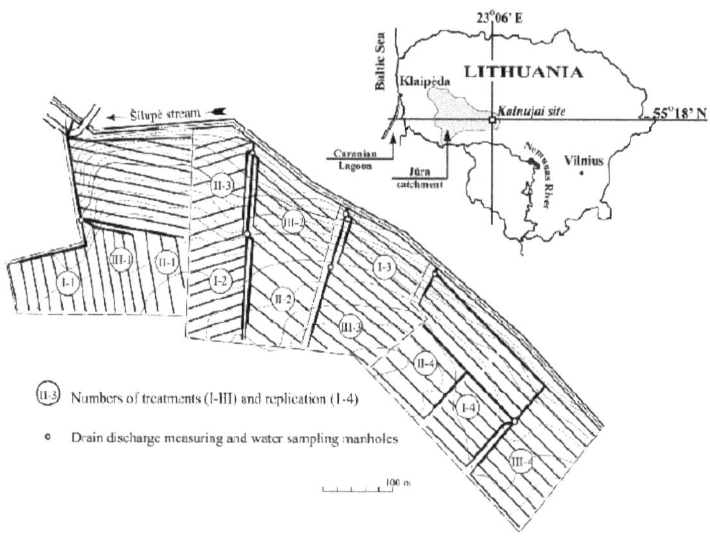

Fig. 2. Geographical location and experimental layout of Kalnujai site.

Drainage water from the main collectors discharges into the modified (deepened and straightened) reach of the Šilupė stream. This stream (length – 4.4 km, basin area – 4.0 km2) according to the Lithuanian river classification, is attributed to small rivers that constitute 75.5% of the total length of all rivers (Gailiušis et al., 2001). Small and medium rivers that usually are drainage water recipients are very important from the ecological point of view because concentrations of biogenic substances there are often higher than in the large rivers.

Experimental drainage systems with lime backfill were installed in 1989 (Fig. 3). The site was drained by composite subsurface drainage systems. Tile drainpipes (length of 33 cm, laterals Ø50 mm, collectors Ø75 mm) have been installed at the depth of 1.1 m, width of drainage trench – 0.5 m. Three drainage treatments have been installed with four replications of each:

1. Drainage trench backfill mixed with lime (0.6% CaO), drain spacing L = 16 m;
2. The same, L = 24 m;
3. (control) – drainage trench backfilled with native clay loam soil, L = 16 m.

Fig. 3. Installation of lime filter drainage in Kalnujai site

Calculations of the optimal lime rate. Laboratory experiments with different clay soil samples were carried out before the lime filter drainage installation in order to determine the optimal rates of lime additives to be mixed with clay soil of a trench backfill. Equation (2) was obtained from the results of laboratory analyses (Blažys et al., 1993):

$$a = 0.13 + 0.011\ N, \quad (2)$$

where a – optimal lime rate considering active CaO and MgO, %; N – amount of physical clay particles contained in soil, %.

This equation should be applied when the soil contains from 20 to 80% of physical clay particles (<0.01 mm). This particular composition was named as physical clay according to soil textural classification by Kačinskij, which was used in the former Soviet Union.

Laboratory findings showed, that water permeability of clay soils is increasing when lime rate is increased to a certain limit that depends on the amount of physical clay particles obtained in the soil. Lime activity should not be less than 70-85% for dry mass as this is the best material to improve the texture of clay soils. The optimal amount of lime that needs to be mixed with clay soil in order to enhance its best hydraulic conductivity may be calculated from the following equation:

$$A = \gamma\ b\ h\ a\ n^{-1}, \quad (3)$$

where A – amount of lime for one linear meter of the trench, kg; γ – soil bulk density, g/cm3; b – width of drainage trench, m; h – depth of drainage trench, m; n – neutralizing value of lime according to the total amount of CaO and MgO, %; a – optimal amount of lime required to improve soil permeability, %.

Silty clay soil samples (sum of <0.01 mm physical clay particles made up 73.8%) were mixed with lime additives (0.6% of active lime for soil mass). After 28 days a microphoto was taken and then enlarged 400 times (Fig. 4). The photo clearly shows 15-20% higher soil porosity (black color) than that of the soil where no lime was added. Based on the laboratory results the amount of lime for one linear meter of the trench backfill was calculated out of the soil characteristic and drainage parameters in Kalnujai site (γ = 1.58 g/cm3, b = 0.5 m, h = 1.1 m, N = 43.0%, n = 21.5%). Estimated amount of shale ashes was 24 kg/m.

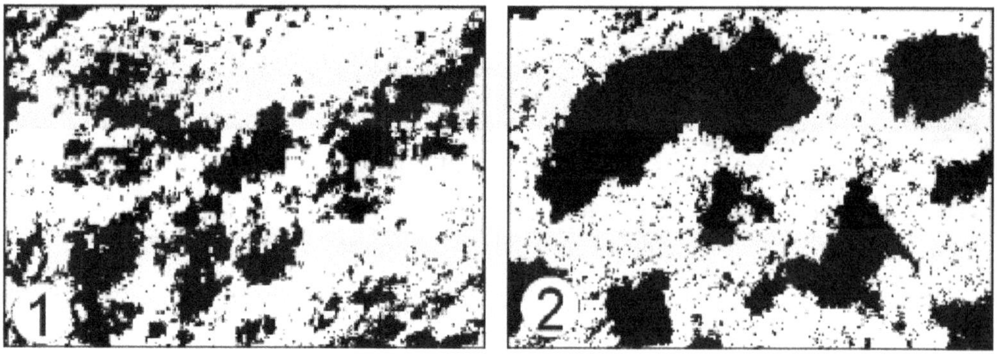

Fig. 4. Microphoto of clay soils (enlarged 400 times): 1 – silty clay; 2 – the same clay mixed with 0.6% CaO

Soil and climate data. The Jūra catchment drains middle part of Žemaičiai Upland, the surface of which is covered with humus rich but poor in labile phosphorus clay loams (80% of the basin area). Almost half (47%) of average annual precipitation amount turns to runoff. Such conditions stimulate soil outwash and inflow of adsorbed phosphorus. High concentrations of suspended matter in the Jūra and its tributaries show the phosphorus loads to be formed there by soil erosion. It must be noted, that Lithuanian soils have low phosphorus retention potential (the definition is based on global soil climate map and global soil map).

Orthi-Haplic Luvisols (*LVh-or*) and *Hapli-Epihypogleyic Luvisols* (*LVg-p-w-ha*) with soaking features prevail in the site. The soils of sandy clay loam/clay loam are maintained at about pH 6.9-7.5. The layer below 40 cm consisted from over 60 to 34% of clay. At the beginning of the investigations the average bulk density determined in the soil profile to the depth of 1.0 m was 1.65±0.02 g/cm3, particle density – 2.67±0.01 g/cm3, porosity – 38.05±0.43%, phosphorus content in the plough-layer (0–23 cm) ranges from very low to medium (27-144 mg/kg), potassium content varied from 84 to 124 mg/kg.

Perennial grasses for hay were grown in the period of four years (1999-2002). In autumn of 2002, the plot was tilled and since 2003 it has been used under crop cultivation. Principal crops are cereals. The data of Meteorological Station located 5 km from the experimental site was used to characterize meteorological conditions. The mean air temperature of the study period was 7.1°C and exceeded the seasonal norm (5.9°C) by about +1.2°C. In the warmest years – 2000, 2002 and 2008 – it was by about +2°C higher than the norm (higher temperatures were more frequent in winter–spring months – December–April and July) (Fig. 5).

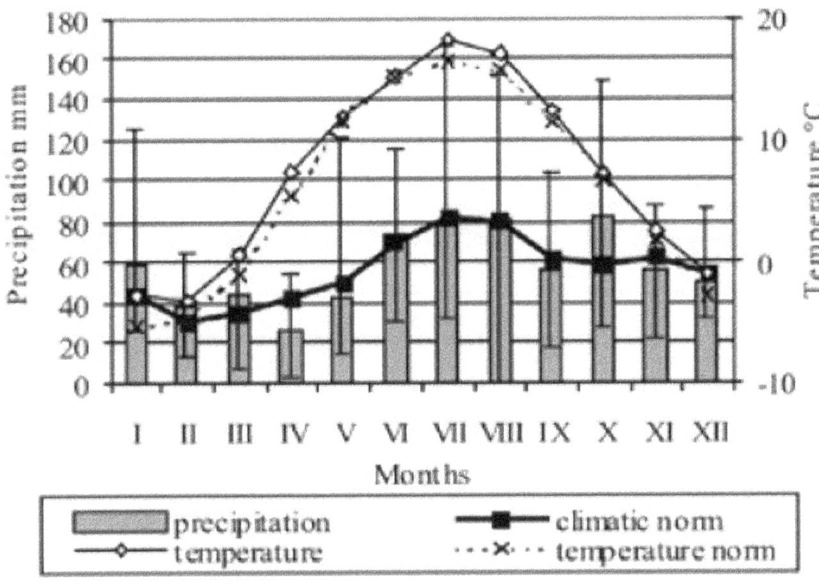

Fig. 5. The means of monthly precipitation and air temperature, 1999–2009 (error bars show data range)

The territory experienced 71% of annual precipitation during the warm season (April – October). Comparison with seasonal norm of annual precipitation the years of 2001 and 2007 attributes to humid ones (precipitation likelihood 14 and 2%). In the dry year (2005) precipitation likelihood was 93%. The remaining period may be considered as moderate: annual precipitation was close to long-term average for this region (665 mm). Significantly uneven rainfall distribution was observed in April and May (variation coefficient V = 68– 73%). In some years precipitation in January, May and October made 2.5–3 of monthly norm, while the average of April made only 60%. Drainage occurred during the entire year of 2001 and 2007, in 2005 dormant period of the drainage lasted even for ten months.

Data collection and analysis. To evaluate the efficiency of lime filter drainage, drain discharge and watertable depth were measured once a day in the main drainage season (from January to May). Drain discharges were measured by volumetric way in the tile outlets, while the depth of the watertable was measured in observation wells installed at the mid-point between the parallel laterals in each treatment. Hydraulic conductivity of drainage trench backfill, topsoil layer and soil between the drains were measured by the same ring infiltration method as in the Pasvalys site.

Monitoring on water quality in the site has been ongoing since 1999. Water samples were analysed for pH, calcium (Ca^{2+}), magnesium (Mg^{2+}), potassium (K+), sodium (Na+), chloride (Cl–), sulphate (SO_4^{2-}), bicarbonate (HCO_{3-}) phosphate (PO_4^{3-}), nitrate (NO_{3-}) and ammonia (NH4) ions also for total phosphorus (TP) and total nitrogen (TN). Periodicity of sampling was once a month. Water samples from the Šilupė stream were taken simultaneously. The methods used to determine chemical parameters are summarized in Table 1.

Table 1. Analytical methods for determining chemical composition of drainage water.

Parameter	Method	References
pH	Electrometry	ISO 10523-1:1994
Calcium	EDTA Titrimetry	LAND 68:2002
Magnesium	EDTA Titrimetry	LAND 73:2005
Potassium	Flame Photometry	LST ISO 9964-3:1998
Sodium	Flame Photometry	LST ISO 9964-3:1998
Chloride	Mohr's method	LAND 63-2004
Sulphate	Gravimetric	ISO 9280
Bicarbonate	Potentiometric	LST EN ISO 9963-2:1999
Phosphate	Spectrophotometry	LAND 58:2003
Nitrate	Spectrophotometry	LST EN ISO 13395:2000
Ammonia	Spectrophotometry	LAND 38:2000

The comparison of ion concentration means during the investigation period was made, i.e. a hypothesis was checked if mixing lime into drainage trench backfill had impact on the basic ions concentrations in drainage water. In parallel, water quality differences of the control drainage treatment and stream water were compared.

The reliability of the results was determined by processing them with mathematical statistical methods, using MS Excel 2000 Data Analysis Tool Pack. Differences of drainage treatments were tested at the significance level $p < 0.05$ and $p < 0.01$. The temporal changes in concentrations of basic ions were evaluated choosing the most reliable trendlines for the data sets. Correlation between phosphorus concentrations in the drainage treatments and meteorological indices (precipitation and temperature) was established. The annual load of phosphorus and nitrogen was calculated on the basis of the linear interpolation method recommended by the Helsinki Commission (Guidelines..., 1994).

3. RESULTS

3.1 The effect of lime admixture to trench backfill on the functioning of tile drainage in clay soils

The drainage treatments efficiency was assessed by comparing drainage coefficients in the Pasvalys site. The data obtained shows that the drainage coefficient (q) from the control trench backfill reduces by 13 times on the average when the soil moisture content increases from 23.0 to 27.0% (Fig. 6). When the drainage trench is mixed with shale ashes, lime reduces clay swelling and ensures stabilized clay soil structure. This results in more stable drainage coefficients. When trench backfill is mixed with 0.15 and 0.30% CaO, q decrease by 4.2 and 3.3 times respectively; when 0.40 and 0.80% CaO is applied, – only 2.6 times.

In case when the soil moisture content is less (23.0%), the average drainage coefficients from trench backfill mixed with 0.40% CaO are 0.58 m/d, i. e. 11 times higher than in the control treatment (0.05 m/d). With higher soil moisture content (27.0%) q is even 55 times higher in treatment with 0.40% CaO (0.22 m/d) than in the control treatment (0.004 m/d). This is a rather favourable index for drainage functioning at critical moments.

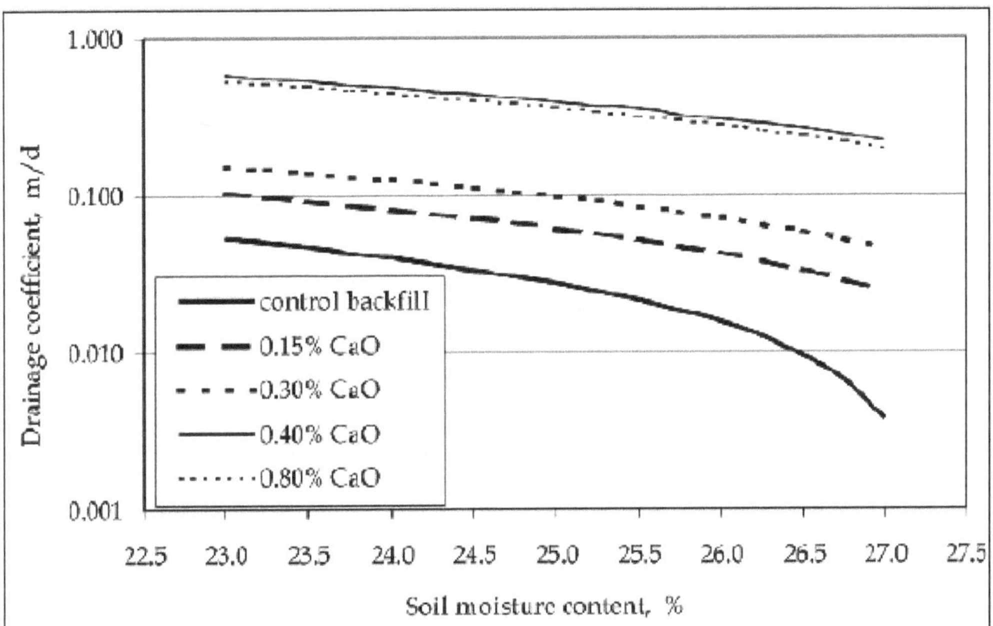

Fig. 6. Trendlines of drainage coefficients in the Pasvalys site.

In case when soil moisture content is higher (27.0%) drainage coefficients from trench backfill mixed with 0.80% CaO are 0.20 m/d, i.e. 50 times higher than in the treatment without lime (0.004 m/d). When trench backfill contained half as much lime (0.40% CaO), q is by 55 times higher. This shows that a certain limit was achieved above which the increased amount of lime additives does not result in the increased permeability of the trench backfill. It could be explained by the large amount of ballast elements contained in shale ashes that frequently make up 85% of the whole material. The ballast elements fill in a certain part of soil pores, and so decrease water permeability of the trench backfill.

The following logarithmic relationship between soil moisture content and drainage coefficients of different drainage trench backfill was determined when analyzing the data over a ten-year period (1987-1997):

$$q = A - B \ln w, \qquad (4)$$

where q – drainage coefficient, m/d; A and B – coefficients of regression equation (Table 2), w – soil moisture content at a 25-30 cm deep soil layer, %.

The equation (4) is valid when the soil moisture content is changing within the range of 23 to 27%. Observations have shown that the changing moisture content of clay soils enhances the changes in water permeability of the trench backfill: with the increase in soil moisture content the drainage coefficients decrease (r = 0.67 - 0.96). A strong reverse relationship was determined between the soil moisture content and drainage coefficients from the control drains and in treatments with 0.40 and 0.80% CaO; a moderate relationship was determined when the drains were backfilled with fewer amounts of lime (0.15 and 0.30% CaO) (Šaulys, 1999).

Table 2. Coefficients of regression equation (4), and comparable values of drainage coefficient (p < 0.01).

No	Type of backfill	Coefficients of regression equation		Drainage coefficient, m/d		Correlation coefficient
		A	B	max	min	
1	Control backfill	1.02	0.31	0.05	0.005	0.96
2	0.15% CaO	1.62	0.48	0.10	0.012	0.77
3	0.30% CaO	2.22	0.66	0.17	0.019	0.67
4	0.40% CaO	7.53	2.22	0.69	0.252	0.82
5	0.80% CaO	7.18	2.12	0.59	0.171	0.81

Analysing the permeability of other drainage backfills it was defined that the permeability of all of them was higher than of the control one. It was substantiated on the calculated discharges per unit length. The data show that with the higher moisture content (27%) in clay soils, discharges from the drains with wood chips backfill are 5 times higher than from the drains with control backfill, with chopped straw backfill – 10, with sand and gravel mixture – 14, with turf backfill – 18, with breakstone backfill – 24 times higher (Fig. 7). The drainage works even more effectively when backfill was mixed with lime (0.40% of active CaO) – 45 times better than drains with control backfill (drainage discharge 0.045 l/s×m).

The obvious advantage of lime filter drainage over other backfills shows the unquestionable efficiency of this measure when draining the surface water excess in heavy clay soils. Since this conclusion was based on the experiments carried out in the model site it was important to investigate efficiency of lime filter drainage under natural field conditions. Further investigations were carried out in the Kalnujai site.

Under the meteorological conditions of Lithuania, drainage outflow commonly lasts till the beginning of May. In autumn it starts again in November. During the months from January to April there are usually several peaks of drainage outflow, the time and duration of which depend on the fluctuations of air temperature and precipitation.

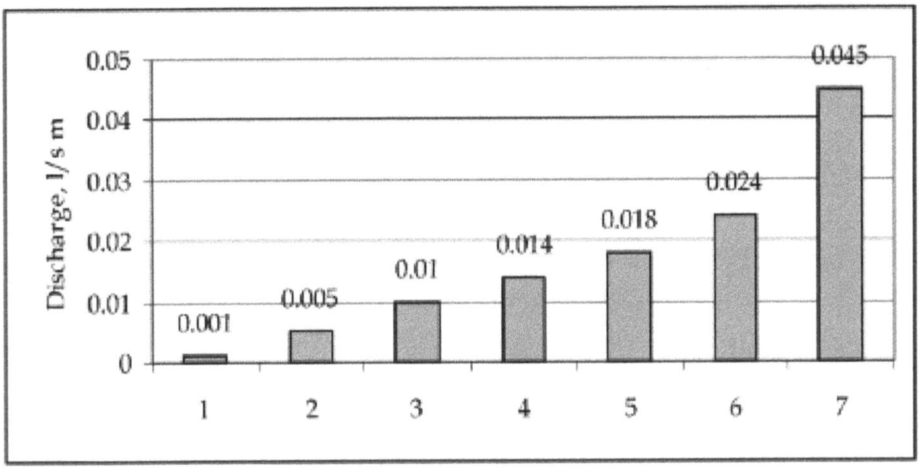

Fig. 7. Comparison of different backfills efficiency in the reference of drainage water discharges per unit length (soil moisture content W=27%): 1 – control (native clay soil) , 2 – wood chips , 3 – chopped straw , 4 – sand/gravel mixture; 5 – turf; 6 – breakstone; 7 – the backfill mixed with lime (0.40 % of active CaO).

Three peaks of drainage discharge were observed in Kalnujai site. The first peak was observed in the first ten-day period of January when the soil surface was frozen. Average drainage discharges were 3.0 mm/d (control), treatments with lime - 3.2–3.3 mm/d. The differences between the control drainage and LFD were 6.6–10%. The highest drainage discharge measured in the system of control drainage during the snow melt period at the outset of February reached 6.7 mm/d and was 1.3 times higher than the design drainage coefficient (5.2 mm/d). At the same time drainage discharge in LFD was lower, reaching 6.2 mm/d in treatment I, and 5.0 mm/d in treatment II. The third slightly lower discharge peak occurred on the middle of March. During this peak drainage discharge in the control system and in LFD I with the same spacing (L=16 m) were equal (4.2 mm/d). In LFD II with spacing L=24 m, drainage discharge was 3.8 mm/d, i.e. decreased by 11%. As the data shows, it is difficult to evaluate the advantage of drainage variants because of great impact of meteorological conditions on drainage work.

The differences between the control and lime filter drainage are more noticeable according to watertable depth between the drains (Fig. 8).

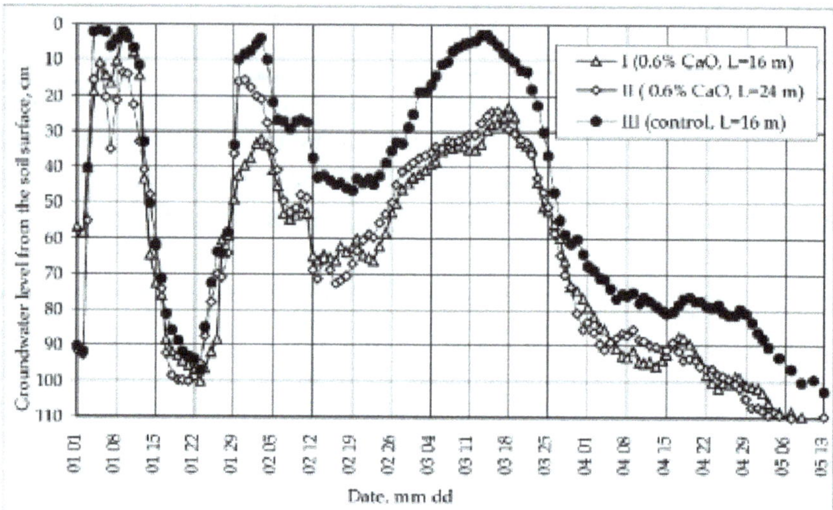

Fig. 8. Dynamic of groundwater level in different drainage treatments (Kalnujai site, 2000) (Šaulys & Bastienė, 2003).

During the main drainage season (from January to April) average depth of the watertable in the control drainage system was 0.16 m (35%) higher than in LFD. Increased drain spacing from 16 to 24 m has no negative impact as the depth of the watertable in LFD varied from 0.003 to 0.025 m and was statistically insignificant (Table 3).

Table 3. Average watertable depth between the drains (m) in 2000.

Treatments	Months				Average (I-IV)
	I	II	III	IV	
I (0.6% CaO, L=16 m)	0.55	0.54	0.42	0.93	0.61
II (0.6% CaO, L=24 m)	0.57	0.52	0.41	0.93	0.61
III (control, L=16 m)	0.49	0.33	0.22	0.77	0.45

The relationship between drainage coefficients and watertable depth in between the drains (Fig. 9) may be expressed by the exponential regression (r = 0.85 - 0.91):

$$q = m\, e^{-nh}, \quad (5)$$

where q – drainage coefficient, m/d; m and n – coefficients of regression equation; h – depth of the watertable in between the drains, m.

The statistical analysis shows there are significant differences between the mean values of watertable depths in the control drainage system and those of the treatments with lime mixed into the trench backfill (Table 4).

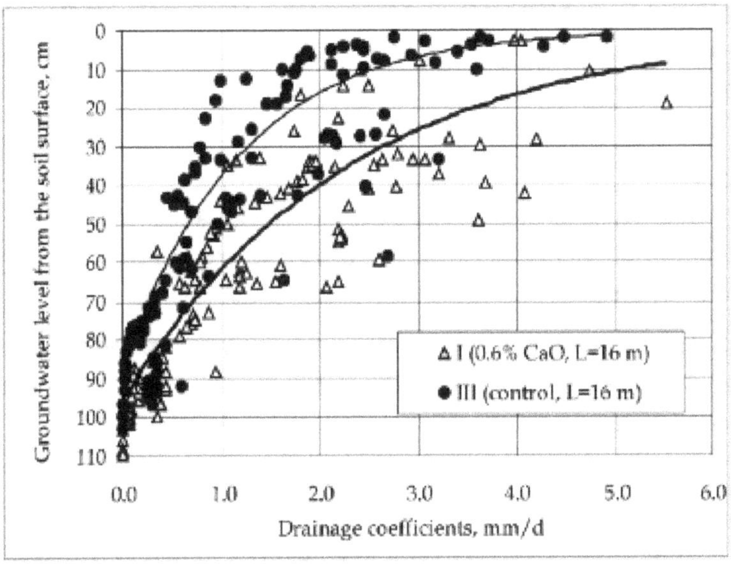

Fig. 9. Relationship between drainage discharge and watertable depth in between the drains (Šaulys & Bastienė, 2003)

Table 4. Statistical estimation of differences in the reference of watertable depth in lime filter drainage and control drainage ($p<0.05$)

Drainage treatments	Number of measurements	Groundwater level differences, m			Standard deviation	Variation coefficient
	n	Max	Min	Mean	SD, m	V, %
2000 (precipitation 112% of seasonal norm)						
I (L=16 m)	60	0.31	0.08	0.20±0.02	0.06	28.6
II (L=24 m)	60	0.29	0.06	0.18±0.02	0.05	30.7
2001 (precipitation 116% of seasonal norm)						
I (L=16 m)	41	0.40	0.07	0.23±0.04	0.08	38.1
II (L=24 m)	41	0.36	0.02	0.23±0.04	0.09	39.4
2002 (precipitation 143% of seasonal norm)						
I (L=16 m)	56	0.62	0.10	0.33±0.05	0.14	43.9
II (L=24 m)	56	0.44	0.08	0.24±0.04	0.11	44.4
2000-2002						
I (L=16 m)	157	0.62	0.07	0.25±0.04	0.10	38.7
II (L=24 m)	157	0.44	0.02	0.21±0.03	0.07	32.5

Note: Mean value ± confidence interval

In the period 2000-2002 the watertable depth in LFD I was on average 0.25±0.04 m deeper than in the control drainage system while in LFD II it was 0.21±0.03 m deeper. It was also observed that during the wet period (in 2002) the differences between watertable depth in LFD and the control drainage were more distinct (0.33±0.05 m and 0.24±0.04 m respectively) than during the dry period (in 2000) (0.20±0.02 and 0.18±0.02 m respectively).

The data obtained at the Kalnujai site were similar to the study results gained by Gurklys (1998), who has determined that watertable between the drains was 20-30% lower when lime was added into the drainage trench backfill (this difference is most obvious when watertable are at their highest level).

According to measurements watertable depth at 0.10 m below the soil surface in the control drainage system lasted for 11 days; at the depths of 0.20 and 0.30 m it remained for 20.5 and 24 days. In drainage treatments with lime it rose only to 0.30 m from the soil surface and remained here for 5.5 and 6 days respectively. The increased drain spacing did not result in less drainage functioning.

Clay soils often have relatively high permeable top layer, which is determined by the agriculture activity. Under extreme conditions topsoil layer saturates by surface water and depression curve does not form. In this case, inflow to the drains can be roughly estimated according to the Donat-Slichter formula (Bereslavskii, 2008):

$$q=1,48\ k_a\ h_a, \quad (6)$$

where q – water inflow from the 1 m width strip formed between the drains, m3/s; ka – hydraulic conductivity of top layer, m/s; ha – thickness of the top layer, m.

In cases of perched watertable (when clay soils are swollen) surplus of water interflows the pervious top layer, which has connection with trench backfill more permeable than the original soil. This flow mode will only be evident if the backfill is of high hydraulic conductivity:

$$q=k_t \cdot b_t, \quad (7)$$

where kt – hydraulic conductivity of the drainage backfill, m/s; bt – width of the trench, m. In order to ensure satisfactory functioning of drainage in clay soils, the inflow through drainage trench backfill should be more than the inflow through the topsoil layer. Between the permeability of topsoil layer and trench backfill the following dependency should be:

$$k_t \cdot b_t \geq 1,48\ k_a h_a. \quad (8)$$

The width of drainage trenches dug by multi-scoop excavator – bt = 50 cm, thickness of the topsoil layer in clay soils ha = 15-20 cm. Based on the equation (8) we receive that:

$$k_t \geq 0,45 \div 0,60\ k_a. \quad (9)$$

The tests of infiltration allowed to compare the hydraulic conductivity of the control and lime filter drainage backfill with hydraulic conductivity of undisturbed clay soil at the same depth between the drains (Fig. 10).

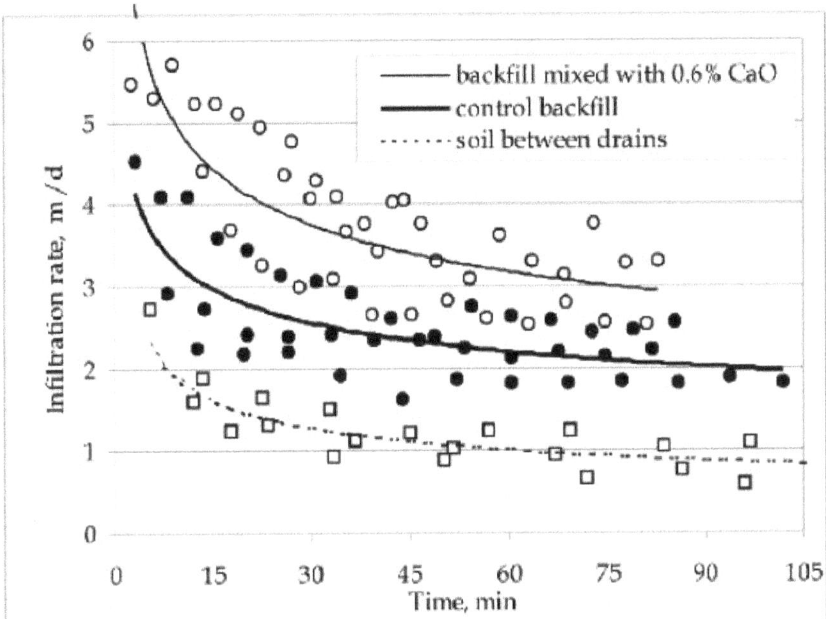

Fig. 10. Variation of water infiltration rate in the Kalnujai site: (o)–trench backfill mixed with lime, (·)–control backfill, (*)–soil between the drains.

The mean value of water infiltration rate in the trench backfill with lime at $p< 0.05$ after the steady flow was established amounted to 3.2±0.25 m/d and was 1.5 times higher than in control backfill – 2.2±0.14 m/d. The water infiltration rate between the drains reached only 0.83±0.09 m/d, i.e. 2.6 times lower than in the control backfill and 3.8 times lower than in the backfill with lime. The power trendlines show the changes of infiltration rate (r = 0.65-0.82).

Lime admixture into the trench backfill in clay soils ensures lower watertables and improved water regime of the body of soil between the trenches because heavy clay soils often have relatively high permeable top layer, which is determined by the agriculture activity. In cases of perched watertable surplus of water interflows the pervious top layer, which has connection with trench backfill more permeable than the original soil. This flow mechanism will only work if the backfill is of high hydraulic conductivity.

3.2 The impact of lime filter drainage on the chemical composition of drainage water

Chemical analysis showed that lime admixture determined changes of ions composition of drainage water (Table 5). Lime admixture has hardly any influence on drainage water pH because the mean values of the pH for the different treatments are almost the same (pH 7.4±0.06). They have relatively small standard deviations and variation coefficients (3.4- 3.7%), pH values of stream water range from 7.4 to 8.4.

Treatment	Parameter	pH	Ca^{2+}	$^*Mg^{2+}$	K^+	$^*Na^+$	Cl^-	SO_4^{2-}	$^*HCO_3^-$	PO_4^{3-}	NO_3^-	NH_4^+
I	Mean	7.4	97.6	38.0	0.95	16.7	16.4	31.2	445.6	0.039	15.5	0.09
(0.6%	Max	8.0	130.0	52.0	2.1	25.0	32.0	60.0	560.0	0.092	58.0	0.88
CaO	Min	6.9	52.0	19.0	0.5	6.0	6.9	14.0	223.0	0.009	0.5	0.001
L=16 m)	$Conf._{0.05}$	0.06	4.0	2.2	0.08	1.3	1.5	2.5	20.2	0.005	3.8	0.04
	CV	3.4	17.1	21.6	33.1	26.8	37.9	32.8	16.0	51.2	130.2	174.2
	n	64	67	48	67	48	67	65	48	67	66	66
II	Mean	7.4	96.4	38.8	0.94	16.8	16.5	31.8	450.7	0.037	16.0	0.08
(0.6%	Max	8.2	127.0	52.0	3.1	29.0	33.0	64.0	595.0	0.098	62.8	0.79
CaO	Min	6.9	50.0	16.0	0.5	5.0	6.0	13.0	209.0	0.003	0.1	0.001
L=24 m)	$Conf._{0.05}$	0.06	4.2	2.5	0.10	1.4	1.4	2.8	23.1	0.005	4.1	0.03
	CV	3.5	18.2	22.7	46.2	30.0	35.6	35.9	17.9	54.9	105.3	177.4
	n	63	66	47	66	47	66	64	47	66	65	66
III	Mean	7.4	99.9	39.2	1.62	17.0	21.5	40.5	440.8	0.114	17.8	0.08
(control)	Max	7.9	135.0	52.0	3.8	23.0	47.0	93.0	560.0	0.193	61.1	0.93
	Min	6.9	53.0	23.0	0.7	9.0	10.0	15.0	240.0	0.031	0.5	0.001
	$Conf._{0.05}$	0.07	4.3	1.9	0.13	0.9	1.7	3.4	19.1	0.007	4.3	0.04
	CV	3.7	18.2	17.2	33.2	19.7	33.4	34.7	15.5	27.2	101.4	179.1
	n	65	68	49	68	49	68	66	49	68	67	68
Stream	Mean	7.9	91.6	21.4	6.42	9.7	19.0	62.0	271.8	0.060	25.5	0.11
	Max	8.4	136.0	39.0	19.0	24.0	35.0	120.0	398.0	0.150	66.4	1.0
	Min	7.4	50.0	10.0	2.0	4.0	8.0	21.0	205.0	0.003	1.8	0.001
	$Conf._{0.05}$	0.04	3.2	1.1	0.66	0.7	0.9	3.9	9.2	0.006	2.8	0.04
	CV	2.8	17.1	21.6	50.8	29.4	23.1	30.7	14.6	52.8	53.3	160.4
	n	90	93	72	93	72	93	91	72	93	93	93

Notes: *observation period of Mg, Na and HCO_3 concentrations - 1999-2004; $Conf._{0.05}$ - confidence interval of calculated mean at 0.05 significance level; CV - variation coefficient in percent; n - number of average concentrations calculated from four replications of water samples.

Table 5. Statistical estimation of basic ions concentrations (mg/l) in the Kalnujai site in 1999-2007.

Calcium and magnesium cations prevail in drainage water. They make 64 and 25% of the total cation amount respectively. In treatments with lime (I and II), the mean Ca2+ concentrations were 97.6±4.0 and 96.4±4.2 mg/l, in the control treatment they were 2–3% higher (99.9±4.3 mg/l). Statistically the differences of these concentrations were insignificant (Table 6). The mean Ca2+ concentration in the stream water was only 5–8% lower compared with the drainage water outflow. However, the estimation of mean differences by t-test showed that it was significant ($p<0.01$). During the entire period of observations Ca2+ concentrations did not exceed maximum admissible concentrations (MAC) in surface water (180 mg/l).

Table 6. Testing of statistical differences between the means of basic ions concentrations in drainage treatments and the Šilupė stream

Treat-ments	Chemical parameters										
	pH	Ca^{2+}	*Mg^{2+}	K^+	*Na^+	Cl^-	SO_4^{2-}	*HCO_3^-	PO_4^{3-}	NO_3^-	NH_4^+
I-III	-	-	-	++	-	++	++	-	++	-	-
III-str	++	++	++	++	++	+	++	++	++	++	-

Notes: - difference statistically insignificant at $p < 0.05$; + difference statistically significant at $p < 0.05$; ++ difference statistically significant at $p < 0.01$.

Average concentrations of Mg2+ in drainage water were 2.5 times lower than the average concentrations of Ca2+. They vary between 16 and 52 mg/l (the mean of I-III treatments – 38.6±2.2 mg/l). At certain moments due to biochemical processes taking place in the soil (especially in summer) the maximum values of Mg2+ concentrations were 30% higher than the MAC (40 mg/l) in surface water. In the stream water average Mg2+ concentration was 21.4±1.1 mg/l.

Migration of K+ depends on the hydrogeological environment of soil (especially on pH). As the pH increases, increased soil cation exchange capacity (CEC) occurs, as well as reduced leaching of basic cations, particularly K+. The research made in the Kalnujai site showed that K+ ions were more stable in the neutral environment. Therefore, it was observed that lime filter drainage water had the average potassium concentrations 1.7 times lower compared with the control treatments and 6.8 times lower compared with the Šilupė stream water (0.95±0.08, 1.62±0.13 and 6.42±0.66 mg/l respectively). The test on the differences between the mean values of the potassium ions concentrations in the different treatments would lead to the conclusion that they are indeed statistically significant at $p < 0.01$.

Analysis show that lime did not have any effect on Na+ concentrations in drainage water: treatments I-II–16.7±1.3 mg/l, III–17.0±0.9 mg/l. Sodium ions concentrations were nearly two times lower in the stream water compared with drainage water (mean value 9.7±0.7 mg/l).

There were significant differences between the mean values of the concentrations of Cl- and SO_4^{2-} anions in drainage water of LFD and in control treatment where they were thereabout 23% higher. Sulphate concentration in the stream water varied from 21 to 120 mg/l (mean value 62±3.9 mg/l). It is noteworthy, that during warm season sulphate concentration in the stream water was higher than it is allowed in open water bodies (MAC=100 mg/l), however, it was not influenced by drainage water but by biochemical processes in the nature.

HCO_{3-} anions prevail in the Kalnujai site water. Their concentrations vary from 209 to 595 mg/l in drainage water and from 205 to 398 mg/l in the stream water. Concentration differences between the drainage treatments are insignificant; values are closely grouped around the mean (CV = 15–18%).

The highest average concentrations of PO_4^{3-} were determined in the drainage outflow of the control treatment (0.114 mg/l). Summing-up all study period the mean value of phosphate ions was 1.9 times higher than the average concentrations in the Šilupė stream water (0.06 mg/l). In the outflow of lime filter drainage PO_4^{3-} concentrations were 66% lower than in the control drainage water. The statistical analysis determined those differences to be significant at $p<0.01$. So, it can be concluded that LFD may reduce phosphate ions concentrations considerably. PO_4^{3-} in Šilupė stream ranged between 0.003 and 0.150 mg/l and do not exceed Lithuanian surface water quality standards.

Water containing nitrate concentrations above 10 mg/l is considered to be of relatively poor quality, whereas 55–60% higher NO3 concentrations were determined in LFD treatments I-II-(15.5±3.8 – 5.9±4.1 mg/l). The average concentrations of NO_3 in control drainage water were 15% higher compared to the treatments with lime. Nitrate levels in the Šilupė stream were about 2.5 times higher and exceeded the guideline concentration for nitrate in surface water.

In the water of small rivers ammonium made up about 5–10% of the total amount of nitrogen, therefore, the concentrations of this ion were much less: in the Šilupė stream – 0.11±0.04 mg/l, in drainage treatments – 0.08±0.03 – 0.09±0.04 mg/l. They did not exceed MAC approved in Lithuania (1 mg/l).

The following regularities were established by analysing the variation tendencies of the basic ions concentrations in the experimental drainage water and the Šilupė stream water during the studies. The fluctuations of Ca2+ ions concentrations were expressed by an Order 2 polynomial trendlines with one valley in the autumn of 2002 (r=0.69–0.76) (Fig. 11). The same type regression was detected when analysing the fluctuations of Mg2+ and Cl- concentrations; however, the trendlines reliability is of less significance (rMg=0.55–0.60, rCl=0.62–0.73).

As the soil solution in the site was near neutrality (pH=6.8) K+ concentrations in LFD water were stably lower compared with the control drainage during the entire observation period. Despite the seasonal variation (CV=33–51%), trends of concentrations of K+ ions generally remained steady throughout the ten-year period both in water of drainage treatments and the Šilupė stream (Fig. 12).

The similar slightly decreasing linear trends of Na+, SO_4^{2-} and HCO_{3-} ions concentrations (rNa=0.31–0.42, rSO4=0.59–0.65, rHCO3=0.32–0.40) in drainage water were detected. At the end of monitoring the mean values of sulphate concentrations decreased by 50% in LFD and by 53% in control treatment in comparison with those observed at the onset of investigations Fig. 13). Accordingly, the mean values of sodium concentrations decreased by 25% (LFD) and by 21% (III) while bicarbonates concentrations – by 20% on the average. At the same time there was no decreasing/

increasing trend in these concentrations in the stream water. They were near the natural background levels for most Lithuanian rivers.

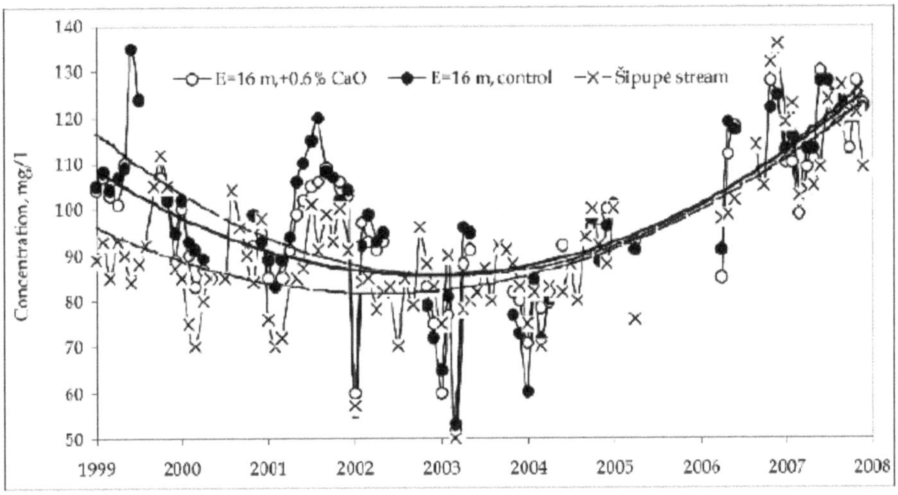

Fig. 11. Variation of calcium cations concentration in drainage treatments and stream water.

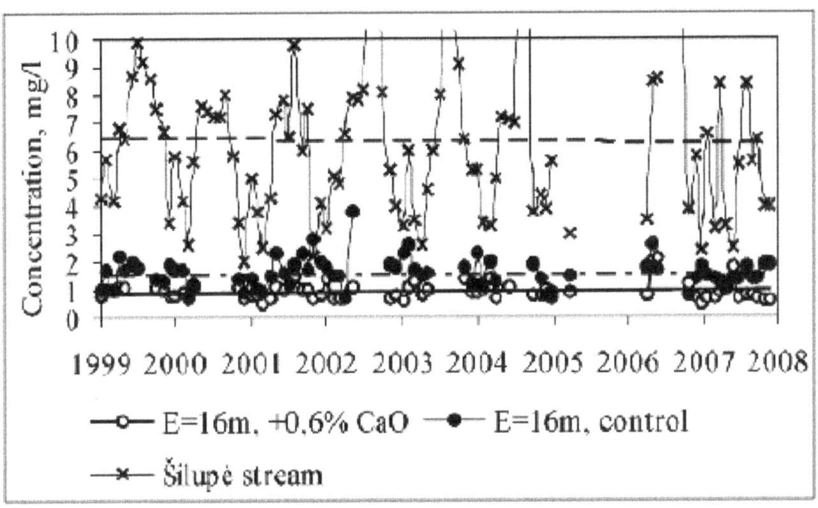

Fig. 12. Variation of potassium cations concentration in the Kalnujai site.

The distribution of PO_4^{3-} concentrations in the stream water and lime filter drainage water showed slightly downward trends (r_{str}=0.32, $r_{(I)}$=0.27). Considerable fluctuations in phosphate concentrations in the control drainage water were detected, therefore, the reliability of chosen linear trendline is very low (^{r}(III)=0.05) (Fig. 14).

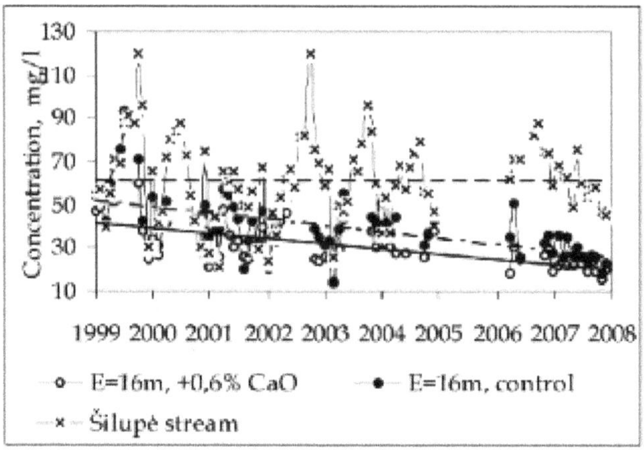

Fig. 13. Dynamics of sulphate anions concentration and temporal trends.

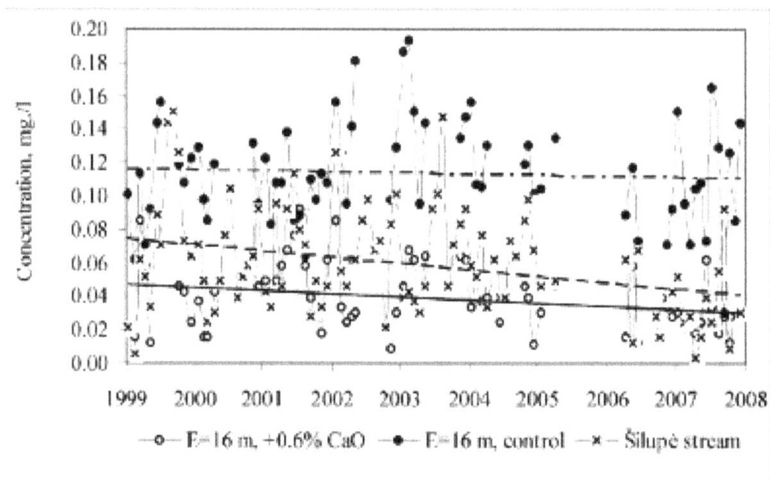

Fig. 14. Temporal changes in phosphate ions concentrations for the period 1999-2008.

The distribution of annual average phosphorus concentration in Šilupė stream water within the period of 1999–2008 shows a slightly decreasing trend of TP (r = 0.49) and $PO_{4-}P$ (r = 0.84). These results confirm the tendencies established earlier and correspond to general character of downward trends found in the Lithuanian rivers within agricultural areas (Povilaitis, 2004; 2006).

Two periods can be discerned when analysing the dynamics of NO_{3-} concentration in subject to land management. First period lasted from the beginning of the investigations until the end of 2002 when the site was used for grassland. The second period spanned since crop cultivation (2003–2007). During the first streak mean values of NO_{3-} in drainage treatments I-III were 0.77, 0.64 and 1.02 mg/l respectively. Conspicuous moderate downward trend (r=0.66–0.74) in all treatments was determined in the time series. The situation changed after the grassland had been tilled in the autumn of 2002. The concentrations of NO_{3-} beganprogressively increase every year. Trends of nitrates of the second streak can be expressed by strong linear regression (r=0.83–0.88). It must be noted that NO_3 concentrations in Šilupė stream water started rising thereabout the year later than in drainage water (from the autumn of 2003).

NH_4^+ level both in the Šilupė stream and drainage water changed widely from the beginning of the investigations until the middle of 2001. Later, fluctuations of ammonium concentrations assumed a moderate character.

3.3 Fluctuation of biogenic substances concentration and loads with drainage outflow

The highest average concentrations of TP and PO4–P were determined in the drainage outflow of the control treatment (Table 7). They reached 0.044±0.006 and 0.032±0.004 mg/l, respectively. Summing-up of all study period the mean value of PO4–P was 1.8 times higher than its average concentrations in the Šilupė stream water (0.018±0.003 mg/l) and 3.2 times higher than its average concentrations in the outflow of LFD treatments (0.010±0.001 mg/l). Extreme annual values of PO4–P (0.049 mg/l) were quantified in the dry year of 2003, when the lowest annual drainage runoff was observed (33 mm).

Table 7. The mean concentrations of phosphorus (mg/l) in drain discharge and stream water in 1999–2009.

Indices	I	II	III	Stream	I	II	III	Stream
	TP				PO4-P			
Mean	0.021	0.021	0.044	0.039	0.010	0.010	0.032	0.018
Max	0.035	0.034	0.060	0.062	0.013	0.015	0.046	0.027
Min	0.012	0.012	0.028	0.026	0.008	0.007	0.022	0.012
SD	0.007	0.007	0.009	0.012	0.001	0.002	0.006	0.005
V %	33	33	20	31	13	22	19	26
Conf 95	0.004	0.004	0.006	0.007	0.001	0.001	0.004	0.003

SD – standard deviation, V % - variation coefficient, Conf95 – 95% confidence interval for a mean TP concentration in the Šilupė steam varied between 0.026 and 0.062 mg/l (the mean value 0.039±0.007 mg/l). PO4–P concentration in receiving stream increased from the lowest values in spring (0.012 mg/l) to 0.027 mg/l in August (the mean value 0.018±0.003 mg/l). According to the rates approved in Lithuania, surface water quality is considered to be good than TP < 0.1 mg/l and PO4–P < 0.065 mg/l. Consequently, the water of the Šilupė stream can be considered as uncontaminated with phosphorus because the concentrations do not exceed

Lithuanian surface water quality standards. However, critical levels of phosphorus in water, above which eutrophication is likely to be triggered, are approximately 0.03 mg/l of dissolved phosphorus and 0.1 mg/l of total phosphorus. According to the UN ECE classification, surface water is considered fairly eutrophic at 0.025 mg P/l consequently the risk of eutrophication in the Šilupė stream still exists. Whereas phosphorus concentrations in the Šilupė stream besides the Kalnujai site were 1.5 times higher than those found in the effluent of LFD, it must be concluded that this drainage practice may reduce surface water pollution with phosphorus compounds.

In Fig. 15 the dynamics of nitrogen concentrations in the Kalnujai site is represented. Despite the seasonal changes, the average concentrations of NO3–N in the Šilupė stream (5.43±0.72 mg/l) generally remained steady during 1999-2003. NO3–N concentrations in the stream water started rising from the autumn of 2003 (approximately the year later than in drainage water). The average NO3–N concentrations of 9.83±1.92 mg/l were determined in 2004-2005.

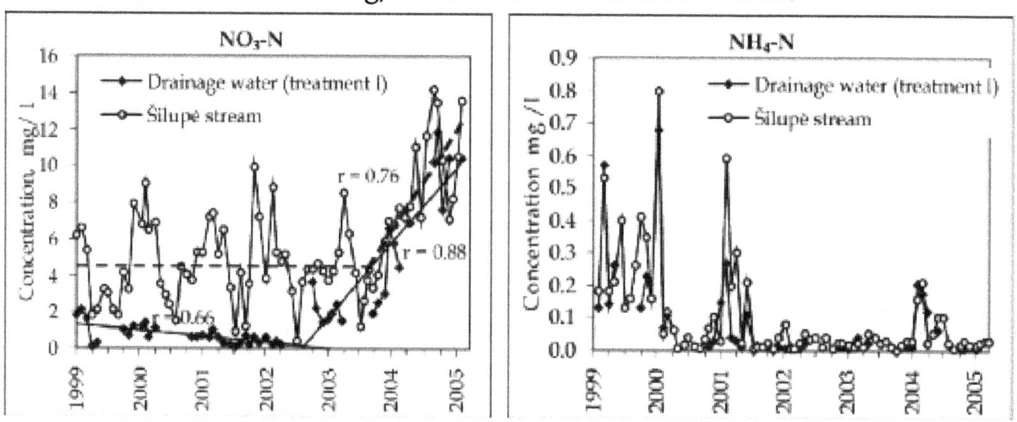

Fig. 15. Dynamics of nitrogen concentrations in the Kalnujai site (dash and line show linear trends of NO3-N in stream water: r = 0.003 and r = 0.76, solid line – in drainage water: r = 0.66 and r = 0.88).

In Lithuanian rivers with natural background level pollution the average concentrations of 0.29 mg/l NO3–N and 0.13 mg/l NH4–N were established (Povilaitis, 2006). The concentrations of NO3–N below 0.3 mg/l are considered to be natural or background levels for most European rivers (Nixon, 2004). It is obvious that in comparison with rivers of natural background level pollution nitrate nitrogen levels in the Šilupė stream are about 20- 30 times higher and exceed the guideline concentration for NO3–N (2.3 mg/l) given for the surface water. The concentrations of NO3–N above 7.5 mg/l are considered to be related with relatively poor quality, though twice-higher concentrations (15.20 mg/l) were determined in 2005.

The annual loads of biogenic substances with drainage outflow in the Kalnujai site are given in Table 8. The greatest annual mean of TP losses (0.06±0.017 kg/ha) was calculated in the control treatment. In LFD the mean leached amounts of TP (0.029-0.031 kg/ha) are 1.9-2.0 times less than in the control treatment; while PO4–P amounts (0.015-0.016 kg/ha) are even 2.7-3.0 times less (0.045±0.011 kg/ha) (the differences are statistically significant at $p < 0.05$). The same differences in terms of total phosphorus loads are statistically significant too. The comparison of phosphorus loads in the drainage treatments with drain spacing L=16 m and L=24 m revealed no significant difference. Hence the lime as amendment of the fine- textured soils positively affects the quality of drainage water and reduces the transport of phosphorus into open water bodies.

Table 8. Annual load of biogenic substances (kg/ha) by drainage outflow in the Kalnujai site in 1999-2005 (p<0.05).

Year	PO4–P			TP			NO3–N		
	I	II	III	I	II	III	I	II	III
1999	0.016	0.014	0.042	0.030	0.029	0.062	1.784	1.585	1.951
2000	0.018	0.018	0.058	0.042	0.040	0.085	2.058	1.449	1.856
2001	0.021	0.026	0.048	0.042	0.047	0.074	5.660	3.690	7.510
2002	0.021	0.017	0.062	0.037	0.033	0.078	0.696	0.491	0.791
2003	0.008	0.008	0.016	0.015	0.012	0.022	13.490	12.320	10.900
2004	0.011	0.018	0.050	0.023	0.034	0.062	8.804	11.590	9.513
2005	0.011	0.012	0.036	0.017	0.021	0.038	9.133	6.652	9.822
Total	0.104	0.114	0.313	0.205	0.216	0.420	41.624	37.777	42.343
Annual	0.015	0.016	0.045	0.029	0.031	0.060	5.946	5.397	6.049
mean*	±0.004	±0.004	±0.011	±0.008	±0.009	±0.017	±3.5	±3.6	±3.2

The negative charge on the clay particles retains ammonium ions (NH4) and protects them from leaching. Nitrate ions (NO3) are negatively charged and are not retained by the clay particles; therefore, subsurface drainage increased the amount of nitrates that can potentially leach from the soil to the drainage water. The calculations showed that in the control drainage treatment NO3–N load amounted to 6.049±3.2 kg/ha per year on the average. In LFD when drain spacing L=16 m the load of nitrate nitrogen amounted to 5.946±3.5 kg/ha per year, when drain spacing L=24 m, NO3–N load was 9% less (5.397±3.6 kg/ha) on the average. During the seven-year investigations the nitrate nitrogen load was 1.7% and 10.8% less from the lime filter drainage systems than that from the control drainage, but these differences could not be treated as statistically significant. Furthermore, it must be noted that nitrates load in particular years differed much more (variation of data 72-91%) than phosphorus, the load variation of which was only 35-38%.

At the same treatment the ratio between marginal values of nitrate load varied from 13.8 to 25.1 during the investigation period. The largest nitrate nitrogen input to the stream water observed in 2003 is likely to be related to intensified agricultural activities (tillage of grassland and crop cultivation, application of mineral fertilizers). Other references also proved that the magnitude of nitrogen loss depended on the soil management (Zucker & Brown, 1998; Povilaitis, 2006). However, the results obtained in the Kalnujai site revealed that the drainage trench backfill permeability improved with lime additives did not increase nitrate leaching to drainage water.

3.4 Seasonal variability of phosphorus concentrations in lime filter drainage outflow

Research shows that the differences in between conventional drainage and LFD have a seasonal variation in respect of phosphorus concentrations in drain discharge (Fig. 16). The lowest concentration of TP was determined in September, whereas the highest one – in January. The values of TP concentrations from the plot with LFD shows more variability depending on seasonal peculiarities (V = 33%) versus phosphate phosphorus (V = 13–22%). Dils & Heathwaite (1999) appeal to field experiments in mixed agricultural catchment in the UK and state that phosphorus concentrations in drain discharge are low (<100 µg TP/1) and stable during base-flow periods (<0.5 1/min), and generally lower than the ones in the receiving stream. In contrast, temporary (hours) elevated P peaks are measured in drain- flow during high discharge periods (>10 1/min). Large sediment-associated particulate P losses are measured during the first major drain-flow event of the autumn.

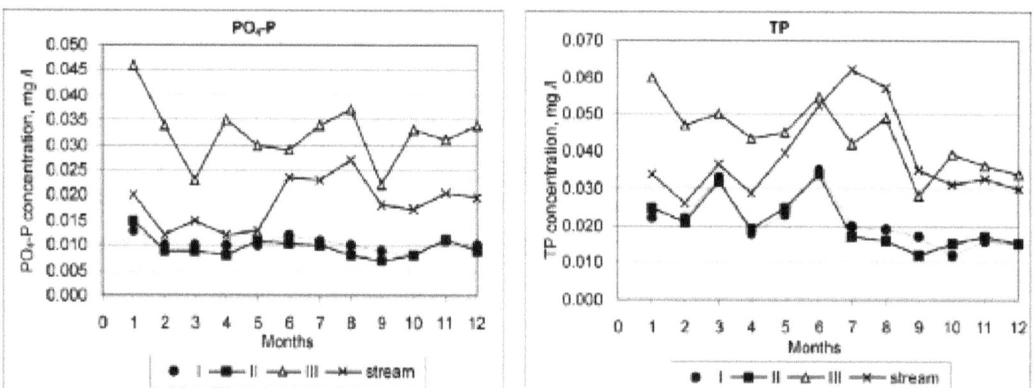

Fig. 16. Seasonal variability of mean phosphorus concentrations in drain discharge and stream water during 1999–2009 (I-II – lime filter drainage, III-control drainage)

Results of studies carried out in the Kalnujai site indicate otherwise. The highest PO4-P concentrations in control drainage outflow are fixed during the entire year with extreme values in January. In the Šilupė stream increased PO4–P concentrations are observed in summer, decreased ones – in spring. However, in all months the PO4–P content in receiving stream is lower than that in the control drainage outflow but exceeds that in the outflow from LFD. This confirms that in cultivated areas drainage has a significant impact on PO4–P content in surface water and such means as LFD in clay soils can reduce this negative impact.

The salient inflow of TP in receiving stream was observed in warm season from April to July later concentrations gradually decrease. The highest phosphorus concentrations (0.052-0.062 mg/l) are observed in summer time (June–August) when runoff is low and intensive release of dissolved phosphorus from the sediments takes place. From July to October is the period when TP content in the stream is higher than that in the drainage outflow. A. Povilaitis (2004) states that trends in the change of phosphorus concentrations in Lithuanian rivers are strongly affected by fluctuations of runoff and phosphorus content in streambed sediment. Taking into account the fact that high quantities of phosphorus can be transported with suspended sediments, the significantly increased loads of suspended matter during the heavy rains in summer can result in the increased particulate phosphorus content.

The changes in predominating phosphorus forms in the Šilupė stream depend on the season of the year. The portions of phosphate phosphorus amounts make from 33% to 63% of total phosphorus amount there. The largest ones were detected in winter months. Thus, the water outflowing from the conventional drainage systems can influence water quality in the recipient stream in cold season only but this influence may vary.

Comparison of lime filter drainage (I and II) and control drainage (III) outflow gives ambiguous results, which show that significant differences develop only in certain months (Table 9).

Table 9. Statistical estimation of phosphorus concentrations in drainage outflow and in the Šilupė stream in the period of 1999–2009 (+ significant difference at *p*<0.05; ++ significant difference at *p*<0.01; – no differences)

Treat-ments	Months 1	2	3	4	5	6	7	8	9	10	11	12
TP												
I-II	–	–	–	–	–	–	–	–	–	–	–	–
I-III	++	+	+	+	+	–	–	–	–	–	++	++
II-III	++	+	+	+	+	–	–	–	–	–	+	++
I-st	+	–	–	–	–	–	++	+	+	++	+	++
II-st	–	–	–	–	–	–	++	++	++	+	+	++
III-st	+	–	–	–	–	–	–	–	–	–	–	–
PO4-P												

Treat-ments	Months											
	1	2	3	4	5	6	7	8	9	10	11	12
I-II	–	–	–	–	–	–	–	–	–	–	–	–
I-III	++	+	++	++	++	–	–	–	–	–	++	++
II-III	+	+	++	++	++	–	–	–	–	–	++	++
I-st	–	–	–	–	–	–	–	–	–	–	++	–
II-st	–	–	–	–	–	–	+	–	+	–	++	+
III-st	+	+	–	+	+	–	–	–	–	–	–	–

Hence, the effectiveness of LFD depends on the meteorological conditions of the season. It has been established that in cold season (November–January) TP concentrations decrease by 56–63% in the outflow of LFD (significant differences at $p<0.01$), differences are less significant in February – May period ($p<0.05$), and no significant differences have been established for the June – October period due to high variation of data. Similar variability detected for PO4–P concentrations. According to the data of wet year, phosphate phosphorus content in LFD treatments is significantly lower than that in conventional drainage ($p<0.01$) in all months, with exception of June – October period, when drainage outflow usually stops.

Comparison of TP concentrations in LFD outflow and in the stream shows that significant differences become distinct in the beginning of July, after the discharge has decreased, and persist until January – in this period TP concentration in the stream increases 2.0–3.6 times. Significant differences between PO4–P concentrations in the stream and in drainage outflow are estimated in November ($p<0.01$), and those at the significance level $p<0.05$ – in July, September and December. In the comparison of LFD no significant differences between phosphorus concentrations have been established.

Differences between phosphorus concentrations in stream water and control drainage outflow are observed in some months only. In January – February TP concentration in drainage outflow is 1.8 times higher than that in the stream, but statistically significant differences are estimated only in January. PO4–P concentration in drainage outflow significantly increases (2.3–2.9 times) in January – May, with exception of March. In any other time of year the PO4–P content in drainage outflow was 1.2–1.5 times higher, than that in the stream water, but statistically differences are insignificant.

In the period of observations phosphorus concentrations in LFD outflow do not pertain neither to precipitation amount nor to fluctuations of air temperature (Fig. 17, Fig. 18). This brings to the statement that the effectiveness of lime does not depend on meteorological conditions.

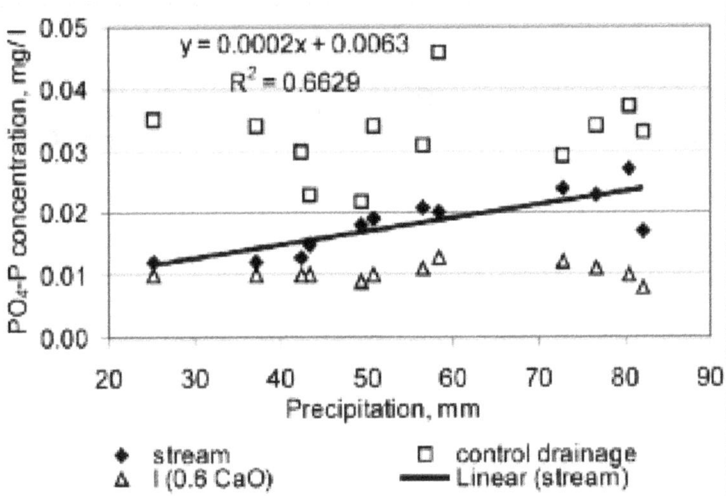

Fig. 17. Correlation between monthly precipitation (x) and mean PO4-P concentration in drain discharge and stream water (y)

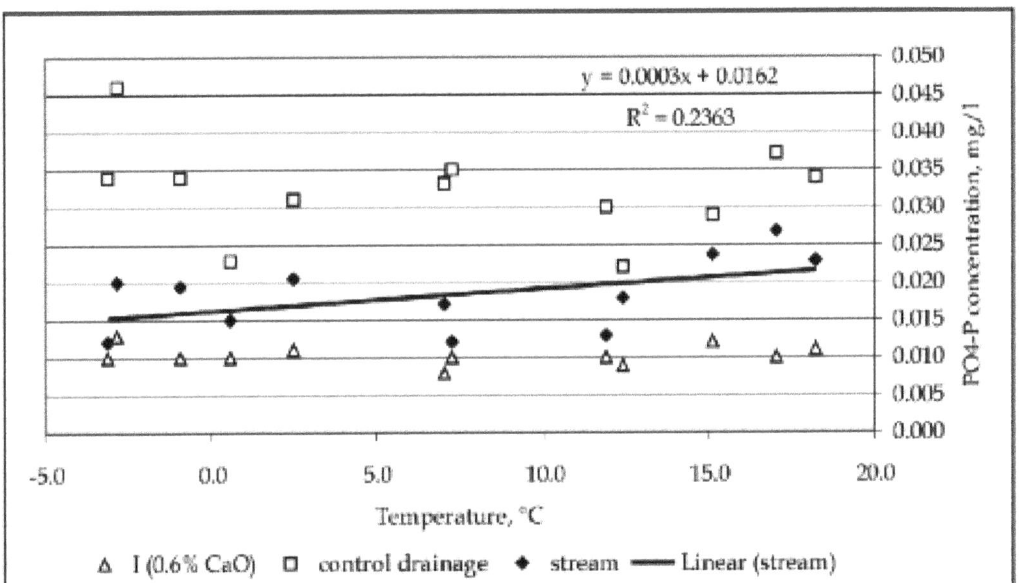

Fig. 18. Correlation between average monthly temperature (x) and mean PO4-P concentration in drain discharge and stream water (y)

Correlation between phosphorus concentrations in the control drainage outflow and the amount of precipitation and temperature are very weak. At the same time strong correlation between TP concentration in surface water and monthly precipitation (r = 0.82), and medium correlation (r = 0.49) between the PO4–P concentration and the temperature was estimated (Table 10).

Table 10. Pirson's correlation coefficients between phosphorus concentration and climate indices.

	I	II	III	Stream	Precipitation mm	Temperature °C
TP						
I	1					
II	**0.827**	1				
III	0.480	0.571	1			
Stream	0.364	0.130	0.211	1		
Precipitation mm	0.058	0.049	0.190	**0.817**	1	
Temperature °C	-0.113	-0.328	-0.236	0.486	0.503	1
PO4-P						
I	1					
II	0.935	1				
III	0.655	0.759	1			
Stream	0.301	0.134	0.257	1		
Precipitation mm	-0.084	-0.124	0.112	0.689	1	
Temperature °C	0.071	-0.106	-0.099	0.805	0.503	1

4. CONCISE SUMMARY

The hydrological functioning of subsurface lime filter drainage systems installed on clay soils is more efficient compared to conventional drainage systems. Incorporating lime into the trench backfill leads to a substantial increase in the pace at which surface water may penetrate the soil, a decrease in the level of the water table between the drains, and improved stability of drainage coefficients during crucial periods.

The implementation of a lime filter drainage system has a beneficial impact on the overall quality of drainage water. This process has the potential to significantly decrease the levels of potassium, chloride, sulphate, and phosphate in drainage water.

The total phosphorus concentrations in the outflow of lime filter drainage were reduced by 50%, while the phosphate phosphorus concentrations were reduced by 64.4% compared to the control drainage water. Thus, it can be definitively established that the addition of lime to trench backfill has the potential to decrease the transit of phosphorus into nearby open water bodies.

Nevertheless, the efficiency of lime filter drainage in Lithuania is influenced by seasonal variations. During the cold season (November-January), the TP concentrations in the outflow of lime filter drainage are estimated to be 2.3-2.7

times lower, with significant differences at a p-value of 0.01. From February to May, the differences are less significant, ranging from 1.5-2.0 times lower with a p-value of 0.05. No significant differences have been observed for the period of June to October. The concentration of PO4–P in lime filter drainage treatments is much lower than in conventional drainage (p=0.01), except during the June-October period when drainage outflow typically ceases.

The results collected from experimental locations in Lithuania demonstrate that lime filter drainage is a durable method for enhancing clay soils. This is evident from the consistent and significant differences noticed in drainage treatments that have been in operation since 1989. Thus, this drainage practice might be considered an efficient method for mitigating non-point pollution of surface waterways in agricultural areas dominated by clay soils.

REFERENCES

Aström, M.; Österholm, P.; Bärlund, I., & Tattari, S. (2007). Hydrochemical effects of surface liming, controlled drainage and lime-filter drainage on boreal acid sulphate soils. *Water, Air & Soil Pollution,* Vol. 179, No. 1-4, pp. 107–116.

Bärlund, I.; Tattari, S.; Yli-Halla, M. & Åström, M. (2005). Measured and simulated effects of sophisticated drainage techniques on groundwater level and runoff hydrochemistry in areas of boreal acid sulphate soils. *Agricultural and Food Science,* Vol. 14, pp. 98–111.

Bereslavskii, E. N. (2008). Application of the Principle of Symmetry to the Solution of the Slichter Problem. *Russian Mathematics,* IZ VUZ, Vol. 52, No. 2, pp. 1–5.

Bergström, L.; Djodjic, F.; Kirchmann, H.; Nilsson, I. & Ulén, B. (2007). Phosphorus from Farmland to Water - Status, Flows and Preventive Measures in a Nordic Perspective. *Report Food 21,* No. 4, SLU.

Blažys, B.; Šaulys V.; Bastienė, N. & Rimas, Š. (1993). Improvement of drainage systems equipment. *Land Reclamation,* Vol. 21, pp. 10-32.

Busman, L. & Sands, G. (2002). *Agricultural drainage.* Publication Series MI-07740.

Cho, H.; Rooij, G. H. & Inoue, M. (2005). The Pressure Head Regime in the Induction Zone During Unstable Nonponding Infiltration. *Vadose Zone Journal,* Vol. 4, pp. 908-914.

Cox, J. W.; Varcoe, J.; Chittleborough, D. J. & Van Leeuwen, J. (2005). Using Gypsum to Reduce Phosphorus in Runoff from Subcatchments in South Australia. *J. Environ. Qual.,* Vol. 34, pp. 2118–2128.

Curtin, D. & Syers, J. K. (2001). Lime-Induced Changes in Indices of Soil Phosphate Availability. *Soil Sci. Soc. Am. J.,* Vol. 65, pp. 147–152.

Dekker, L. W.; Ritsema, C. J. & Oostindie, K. (2001). Preferential Flow in Sand, Loam, Clay, and Peat Soils. *Soil Science: Past, Present and Future* – Joint Meeting of the CSSS and SSSA. Prague, Czech Republic, September 16-20, 2001. pp. 71-72.

Dils, R. M. & Heathwaite, A. L. (1999). The controversial role of tile drainage in phosphorus export from agricultural land. *Wat. Sci. Tech.*, Vol. 39, No. 12, pp. 55–61.

Djodjic, F.; Börling, K. & Bergström, L. (2004). Phosphorus leaching in relation to soil type and soil phosphorus content. *J. Environ.Qual.*, Vol. 33, pp. 678–684.

Doss, B. D.; Dumas, W. T. & Lund, Z. F. (1979). Depth of lime incorporation for correction of subsoil acidity. *Agron. J.* Vol. 71, No. 4, pp. 541-544.

Ekholm, P.; Kallio, K.; Turtola, E.; Rekolainen, S. & Puustinen, M. (1999). Simulation of dissolved phosphorus from cropped and grassed clayey soils in southern Finland. *Agric. Ecosyst. Environ.*, Vol. 72, pp. 271–283.

Forsburg, N. E. (2003). *The Trickle Down Theory: Hydraulic Properties of Soil*. J0609.

Foy, R. H. & Dils, R. (eds). (1998). Practical and innovative measures for the control of agricultural phosphorus losses to water. *Proceedings of an OECD sponsored workshop*, Dep. of Agriculture, Antrim, Northern Ireland 16–19 June 1998.

Fraser, H. & Fleming, R. (2001). *Environmental Benefits of the Tile Drainage*. Ridgetown College– University of Guelph.

Gailiušis, B.; Kovalenkovienė, M. & Jablonskis, J. (2001). *The Lithuanian Rivers*. Hydrography and Runoff. Lithuanian Institute of the Energy, Kaunas.

Guidelines for the third pollution load compilation (PLC-3). (1994). *Baltic Sea environmental proceedings*. Helsinki Commission. Helsinki.

Gurklys, V. (1998). The improvement of soil permeability in drainage backfills. *Water Management Engineering*, Vol. 4, No. 26, pp. 39-46.

Halme, T.; Jaakkola, A.; Kanerva, T.; Horn, R. & Pietola, L. (2003). Effects of ploughpan liming and loosening on soil aeration and root growth. *Proceedings of Nordic Association of Agricultural Scientists 22nd Congress*, Turku, Finland, July 1-4, 2003.

Heckrath, G.; Brookes, P. C.; Poulton, P. R. & Goulding, K. W. T. (1997). Phosphorus losses in drainage water from an arable silty clay loam soil. In: *Phosphorus loss from soil to water*, H. Tunney, O. T. Carton, P.C. Brookes, & A.E. Johnston (Eds.) pp. 367–369. Oxford: CAB International.

Heiskanen, A. S.; Van de Bund, W.; Cardoso, A. C. & Nõges, P. (2004). Towards good ecological status of surface waters in Europe - interpretation and harmonisation of the concept. *Water Science & Technology*, Vol. 49, No. 7, pp. 169–177.

Mubeen, M. M. (2005). Stabilization of soft clay in irrigation projects. *Irrigation and Drainage,*Vol. 54, No. 2, pp. 175-187.

Murphy, P. N. C. & Stevens, R. J. (2010). Lime and Gypsum as Source Measures to Decrease Phosphorus Loss from Soils to Water. *Water Air Soil Pollution,* Vol. 212, No. 1-4, pp. 101–111.

Povilaitis, A. (2004). Phosphorus trends in Lithuanian rivers affected by agricultural non- point pollution. *Environmental Research, Engineering and Management,* Vol. 4, No. 30, pp. 17–27.

Povilaitis, A. (2006). Impact of agriculture decline on nitrogen and phosphorus loads in Lithuanian rivers. *Ekologija (Ecology),* Vol. 1, pp. 32–39.

Puustinen, M. (2001). *Management of runoff water from arable land.* Finnish Environment Institute, Helsinki.

Rajasekaran, G. & Narasimha, R. S. (2002). Permeability characteristics of lime treated marine clay. *Ocean Engineering,* Vol. 29, No. 2, pp. 113–127.

Rhoton, F. E. & Bigham, J. M. (2005). Phosphate Adsorption by Ferrihydrite-Amended Soils. *J Environmental Quality,* Vol. 34, pp. 890–896.

Ritzema, H. P. (ed.). (1994). *Drainage Principles and Application.* The Netherlands. ILRI Publication 16: 1125.

Ryan, M. (1998). Water movement in a structured soil in the south-east of Ireland: preliminary evidence for preferential flow. *Irish Geography,* Vol. 31, No. 2, pp. 124– 137.

Šaulys, V. (1999). The increase of water permeability of the drainage trench backfills in the heavy textured soils. *Water Management Engineering,* Vol. 7, No. 29, pp. 115–126 (in Lithuanian).

Šaulys, V. & Bastienė, N. (2003). Investigations on the filtration of drainage trench backfills. *Water Management Engineering,* Vol. 23-24, No. 43-44, pp. 5–14 (in Lithuanian).

Šaulys, V. & Bastienė, N. (2008). The impact of lime on water quality when draining clay soils. *Ekologija (Ecology),* Vol. 54, No. 1, pp. 22–28.

Šileika, A. S.; Gaigalis, K.; Kutra, G. & Šmitienė, A. (2005). Factors Affecting N and P Losses from Small Catchments (Lithuania). *Environmental monitoring and assessment,* Vol. 102, No. 1-3, pp. 359-374.

Šileika, A. S. (2010). Fyris NP: Cattchment model for assessing phosphorus sources, retention and reduction options in the river Nemunas (Lithuania). *Water Management Engineering,* Vol. 37, No. 57, pp. 34–45.

Skaggs, R. W.; Breve, M. A. & Gilliam, J. W. (1994). Hydrologic and water quality impacts of agricultural drainage. *Critical Reviews in Environmental Science and Technology*, Vol. 24, No. 1, pp. 1-32.

Smedema, L. K; Voltman, W. F. & Rycroft, D. W. (2004). *Modern Land Drainage - Planning, Design and Management of Agricultural Drainage Systems* (second edition), Taylor and Francis, ISBN 90 5809 554 1, The Netherlands.

Tuli, A.; Hopmans, J. W.; Rolston, D. E. & Moldrup, P. (2005). Comparison of Air and Water Permeability between Disturbed and Undisturbed Soils. *Soil Sci Soc. Am. J.*, Vol. 69, pp. 1361-1371.

Tuller, M. & Or, D. (2003). Hydraulic Functions for Swelling Soils: Pore Scale Considerations. *Journal of Hydrology*, Vol. 272, pp. 50-71.

Ulén, B. (2003). Concentrations and transport of different forms of phosphorus during snowmelt runoff from an illite clay soil. *Hydrol. Processes*, Vol. 17, pp. 747–758.

Ulén, B. (2007). Phosphorus from farmland to water in Sweden. In: Abs. of the Workshop *Mitigation options for nutrient reduction in surface water and groundwaters*. COST Action 869, 27-29 November, 2007, North Wyke, Devon, UK.

Valsami-Jones, E. (Ed.). (2004). *Phosphorus in environmental technologies - principles and applications*. IWA Publishing. ISBN: 1 84339001 9.

Van den Eertwegh, G. A. P. H.; Nieber, J. L.; de Louw, P. G. B.; Van Hardeveld, H. A. & Bakkum, R. (2006). Impacts of drainage activities for clay soils on hydrology and solute loads to surface water. *Irrigation and Drainage*, Vol. 55, No. 3, pp. 235–245.

Van der Salm, C.; Van Beek, C. & Van de Weerd, R. (2007). Influence of P-status and hydrology on phosphorous losses to surface waters on dairy farms in the Netherlands. In: Abs. of the Workshop *Mitigation options for nutrient reduction in surface water and groundwaters*. COST Action 869, 27-29 November, 2007, North Wyke, Devon, UK.

Zucker, L. A.; Brown, L. C. (Eds.). (1998). *Agricultural drainage: Water Quality Impacts and Subsurface Drainage Studies in the Midwest*. Ohio State Univ. Extension Bulletin 871.

Weppling, K. & Palko, J. (1994). FOSTOP - a new method to improve water permeability and reduce phosphorus leaching in heavy clay soils. *Agrohydrology and Nutrient Balances*, Uppsala, Sweden, 18–20 Oct., 1994.

Withers, P. J. A. & Haygarth, P. M. (2007). Agriculture, phosphorus and eutrophication: a European perspective. *Soil Use and Management*, Vol. 23, No. 1 (suppl.), pp. 1–4.

10 EXPERIMENTAL VARIANTS AND TECHNICAL EFFICIENCY OF TILE DRAINAGE ON AGRICULTURAL LANDS IN NORTH-EAST ROMANIA

1. INTRODUCTION: AN OVERVIEW

Romania is located in the central part of Europe, namely in the southeastern region of Central Europe. It is situated to the north of the Balkan Peninsula, approximately halfway between the Atlantic Coast and the Urals. Romania is positioned inside and outside the Carpathian Mountains, along the lower course of the Danube River, and has access to the Black Sea. Alternatively, the parallel of latitude 45°N meets with the meridian of longitude 25°E in close proximity to the geographical midpoint of the country, around 100 km northwest of the country's capital, Bucharest. Romania is the eleventh largest country in Europe, with a land area of 238,391 square kilometers.

The relief of Romania is characterized by three primary elevations: the highest being the Carpathian Mountains, the medium elevation corresponding to the Sub-Carpathians, hills, and plateaus, and the lowest elevation consisting of plains, meadows, and the Danube Delta.

Romania has a temperate-continental climate that transitions between several influences. It is influenced by the ocean from the West, the Mediterranean from the South-West, and excessive continental elements from the East. The average temperature varies based on latitude, with a temperature of 8°C in the North and over 11°C in the South. It also varies based on altitude, with a temperature of -2.5°C at lower elevations in the mountains and 11.6°C in the plains.

The annual precipitation decreases in magnitude from west to east, ranging from above 600 mm to less than 500 mm in the East Romanian Plain, below 450 mm in Dobrogea, and approximately 350 mm along the seaside. In hilly regions, the precipitation levels range from 1000-1500 mm.

The rivers of Romania are arranged in a radial pattern, with the majority of them originating from the Carpathian Mountains. The primary watercourse that serves as a collection point for water in the country is the Danube river. It traverses the southern part of the country for a distance of 1075 kilometers before ultimately emptying into the Black Sea (Romanian Statistical Yearbook, 2010).

The maps display the geographical distribution, at both national and district levels, as well as the intensity of three types of excessive moisture: groundwater, rainfall, and flooding. The classification is based on the intensity of the surplus moisture, whereas the subclasses are determined by the nature or source of the moisture (Munteanu et al., 2004).

Drainage systems typically decrease the amount of water that drains from a field by an average of 20-30%. However, the specific amount of water that drains can vary significantly based on factors such as soil type, rainfall, type of drainage system, and management intensity (Ramoska et al., 2011).

The effect of drainage on crop yields varies. Computer simulations conducted over an extended period of time suggest that the average annual increase in yield is less than 5%. However, the increase could be significant in certain years, depending on the variability of annual precipitation and the specific climatic conditions of the location (Cooke et al., 2002, Troeh et al., 2005).

Drainage systems need to efficiently handle flooding, prevent erosion of streambanks, and safeguard water quality (Lukianas et al., 2006). In order to accomplish this, designers need to combine traditional flood control methods for rare, heavy storms with three fundamental stormwater quality control methods for common, light storms: infiltrating runoff into the soil, storing or delaying runoff for later release, and gradually channeling runoff through vegetation (Ritzema et al., 1996, Singh et al., 2006).

Designs for integrated flood control and stormwater quality control must adhere to multiple requirements in engineering, horticulture, aesthetics, functionality, economics, and safety (Townend et al., 2001, Walker et al., 1989).

The preservation and sustainable use of soil resources also involve the targeted application of hydro-ameliorative measures based on the specific limiting factors of soil fertility.

Monitoring conducted in Romania has determined that there are around 4.0 million hectares of farmland with soil resources that are suitable for agriculture. Out of these, around 3.7 million hectares are arable lands that can support competitive agriculture (Dumitru et al., 2004).

According to the 2008 Romanian Statistical Yearbook, the agricultural land categories affected by various factors that limit their productive capacity are as follows: frequent droughts - 7,100 thousand hectares; periodic excess moisture in the soil - 3,800 thousand hectares; soil erosion and landslide - 7,000 thousand hectares; soil compaction - 6,500 thousand hectares; high and moderate acidity - 3,400 thousand hectares; small and very small reserves of humus in soil - 7,500 thousand hectares; low and very low concentration of mobile phosphorus - 6,300 thousand hectares.

Based on the findings of Man et al. (2002), who conducted specialized investigations in Timis, Arad, Bihor, and Maramures counties, it was determined that there was an estimated agricultural area of 958,000 hectares that was damaged by excessive moisture. Drainage was conducted on agricultural lands with excessive wetness, covering a total area of 854,925 hectares. Out of this, 18,159 hectares were drained using tile and plastic drains.

The spatial distribution of ecosystemic territorial units in the north-east of Romania, namely in Suceava County, revealed that 30% to 40% of the land had soil with excessive moisture. This excess moisture extended from south to north and from east to west. Based on the data assessed by Moca et al. in 1988 and 2001, an area of over 100,000 hectares has experienced a consistent presence of excessive wetness, either permanently or temporarily. Hydro-ameliorative improvements, such as open canal drainage systems and subsurface drainage with ceramic and/or plastic pipes, have been implemented on agricultural lands in the extra-Carpathian region of Suceava County to address excess moisture. To effectively manage soil resources with excessive wetness and a high risk of unsuitability for agriculture, the utilization of photogrammetric products such as photographs and digital data files is necessary.

Suceava County is located in the northern region of Romania and is part of the North-Eastern Development Region, along with five other counties. Suceava County, with a total size of 855,350 hectares, is the second largest county in Romania, accounting for 3.6% of the country's land area. Suceava County has a total surface area of 855,350 hectares. Out of this, 349,544 hectares, which is 41% of the total surface, is used for agriculture. The remaining 505,806 hectares, which is 59% of the total surface, is non-agricultural land (Jitareanu et al., 2009).

The agricultural land resources in Suceava County are constrained by both the small size of the area and its production capability. In order to promote the sustainable utilization of soil resources, a range of hydro-ameliorative measures have been implemented over time. The objective of carrying out these activities was to address the soil's agricultural capacity constraints, such as excessive wetness, vulnerability to flooding, soil erosion, unstable slopes, and other related problems.

Some of the initial methods used to address excessive water on agricultural land in Bucovina included the construction of surface water drainage channels between different landowners' parcels, implementing land forming techniques using ridges and furrows, and creating canals for water evacuation. Historical records indicate that in the late 19th and early 20th centuries, the Agricultural Society of Vienna carried out drainage works in the Radauti Depression. These works included subsurface drainage and minor river regularization. A succession of significant tile drainage projects were undertaken in the Radauti Depression, beginning in 1895 and again between 1903 and 1910.

The implementation of drainage and subsurface drainage systems occurred in multiple phases, primarily spanning from 1950 to 1990. During the initial phase, spanning from 1950 to 1975, drainage systems and river regularization with dikes were constructed. During the period from 1976 to 1990, several antiquated systems underwent enhancements and adjustments in size. In addition, new projects were implemented in certain hydrographic basins, which involved the construction of hydro-ameliorative systems incorporating canal drainage and subsurface drainage.

In the post-1990 social and economic landscape of Romania, characterized by a significant decrease in investments, there was also a notable reduction in land improvements. During this stage, only maintenance work and the finalization of existing upgrades were carried out. Some of the specific tasks performed between 1990 and 2010 include maintaining drainage ditch networks, pipe and tile drains, and implementing flood prevention measures.

Between 1960 and 1990, drainage systems were established on 55,100 hectares of land in Suceava County to eliminate excess moisture. Additionally, tile drainage networks were implemented on 26,300 hectares, and measures to prevent flooding were put in place on 7,400 hectares.

The land use composition of Suceava County is remarkably balanced, considering its location within the Carpathian orogeny and the plateaus unit. Agricultural land accounts for 41% of the total, while forests and forest plantations make up 53%.

The class of Luvosols possesses the highest soil resources, accounting for 25.3% of the effectively charted area. Among the various soil types, the faeoziom is notable with a share of 18%.

The soil resources are distributed among five quality groups, with the majority falling into the third class (35%) which represents soils of average fertility, and the fourth class (30%) which represents soils of low fertility.

The primary constraints on soil quality include excessive wetness impacting 185,316 hectares, water erosion and landslides, compaction, acidity, and nutrient deficiencies.

2. SURFACE AND PIPE DRAINAGE SYSTEMS

Natural drainage of the soil is represented by all conditions of terrain, soil and hydrogeology of an area which causes gravity circulation of water located at a point in excess of the land surface or on soil profile (Mirsal et al., 2004).

In terms of hydroameliorative works, the drainage is all that apply to land surface or underground, from which excess water is eliminated.

Surface drainage works is the ensemble of hydro-ameliorative works which removes excess water from rainfall, land surface or accumulated stagnant at the top of the soil profile, above the hard permeable horizons.

From surface drainage category of works are highlighted the following:

- *Land shaping in slope* that is done, usually with a continuous slope of the land surface after one or more inclined planes, to a gutter or nearest canal drainage network. The purpose of levelling land drainage works is to avoid stagnation of rainwater on the surface of local depressions without drainage.
- *Land shaping in the bedding system with ridges and furrows* is the realization of the greatest slope of the terrain of ridges with widths of 15 to 40 min length from 100 to 500 m and cross slope of 1-3%. These strips are ridges separated by shallow channels with large slopes, easily traversed by agricultural machinery.
- *Mole drainage* consists of galleries of 8 to 14 cm diameter, located at a depth of 0.4 to 0.8 m below surface, and is achieved through a special device. The purpose of these underground galleries is to remove excess water from the upper soil profile derived from precipitation.
- *Deep loosening* of soil profile at depth of 0.4 to 0.8 m is achieved either by plowing or scarifying, to increase the permeability of poorly permeable horizons and infiltration of water into deeper layers of soil profile. In addition to the categories of works mentioned above, surface drainage is completed through a hydropower scheme including collection network channels, outlet channels and all related construction.
- *Ridge ditch* issued for the collection and removal of water from precipitation which is in excess of the land surface.
- *Drainage ditch* is a channel that takes discharged water from the network channels and transports it to the nearest outlet channel being carried long distances between them, from 200 to 500 m.
- *Evacuation channel* is the channel that takes drained water from discharged channels and transports it to the nearest outlet higher point channel or direct in the emissary.

- *Emissary (water course)* is a natural watercourse, in which the drainage system discharges its drained water volume either by gravity or by pumping.

Pipe drainage systems consist of pipes that are employed in hydro-ameliorative works to decrease groundwater levels to the necessary depth for the optimal growth of plants, taking into account specific climatic and soil conditions (Mejia et al., 1998). This category of work involves the installation of various sizes of enclosed tubular drains, network channels, and exhaust collection systems, which are occasionally filled with surface drainage.

2.1 The causes of excess moisture on agricultural land

The main water sources and factors that determine the formation of excess soil moisture conditions in Romania are the following:

- *Precipitation* is directly or indirectly, the main source of excess water in the soil for most agricultural land. In natural conditions of various climatic zones are recorded variations of rainfall, both from a calendar year to another, and in the same year from one season to another. Thus, in the same climate area 800-1000 mm annual rainfall can be recorded in rainy years and only 350-450 mm in dry years. However, large amounts of precipitation are recorded in short intervals. Thus, in intervals from 1 to 5 consecutive days, rainfall totals from 50 to 150 mm and sometimes even more. Excess moisture coming from rainwater accumulates in the roots of crops, and sometimes as a layer on top of pedophreatic soil profile. Depending on local natural conditions, excess moisture coming from precipitation is associated with lower slopes and/or local minor unevenness respectively, with clay soils, poorly permeable.
- *High water table* fed by precipitation, infiltration from river sand other sources contribute to excess moisture on the land of meadows, plain sand low terraces. Excess moisture is manifested by increasing the water table, flooding the area of plant roots as well as small swamps in depression areas. On this land, ground water is usually located at depths between 1 and 5 m, which during rainy or high levels in rivers, lakes and other sources increases almost to the ground surface.
- *The heavy rainfall and high ground water level, easy accessional,* fuelled by one or more of the above sources, also contribute to the formation of excess moisture on the clay soils, hardly permeable. Excess moisture in the form of swamps and/or caused by precipitation of pedophreatic layer, respectively, as a high ground water level, the result of association between water sources said.

- *River water* formed by the outpouring of excess moisture and flooding over the sides of valleys without mills. Floods are recorded normally during winter due to snow melting and in summer after heavy rainfall. Duration and frequency of flooding on local conditions vary from one stream to another, very different durations of time. Typically, floods are more frequent and shorter periods on the rivers of Romania and have a longer character and are less frequently in the Danube meadow. However, low lands are flooded from surface water runoff from higher areas of the surrounding land.
- *The landscape* favours the formation of excess moisture in the soil so the small slopes of the terrain and the local minor unevenness elements which make the water flow and natural drainage of the soil characteristic. These areas are specific areas of meadow and low plains and sometimes on the terraces. Excessive wetting of the soil favours swampiness of the land, especially in humid and subhumid climates.
- *The illuvial clay soil with low permeability* from wet climate and sub-humid areas are also challenging conditions for gravity circulation of water in the profile. The presence of textural horizon, illuvial clay, hard permeable, at a depth of 60-80 cm from ground surface contributes to the accumulation of excess water in the upper horizons of the soil profile.

2.2 Soil water balance

Agricultural land drainage fitting sizes, in general, for non-permanent or permanent regime, depending on soil water balance, respectively, of the elements considered for calculation of specific flow discharged through drains network.

Depending on the causes of formation of excess moisture in the soil which were previously presented, the most representative and most common cases encountered in practice of drainage work are.

- *Temporarily excess moisture with stagnant nature, caused by precipitation*, is specific to farm land with small slopes and local unevenness in humid and subhumid climates with clayey soils hardly permeable. The water balance during periods of excess moisture has the following form:

$$P - ET > S + I + W_{max}$$

where:

P - average annual rainfall;

$P - ET > S + I + Wmax$

ET - average annual evapotranspiration; S - surface water drainage of land;

I - the amount of water infiltrated in the soil profile, below the plant roots;

W_{max} - the maximum amount of water that can accumulate in the soil in the plants root zone.

In practical terms, this formula in a simplified form of soil water balance equation is complete in terms of knowledge of natural factors causing excess moisture in the soil and established necessary drainage work.

In this case, **the excess water volume** (**Ve**) can be expressed by the relationship:

$$Ve = P - Et - S - I - W_{max}$$

Specific agricultural land of humid and sub-humid climates are located on terraces, high and piedmont plains, highlands and hills: Banat, Crisana, Maramures, Transylvania and the northern Oltenia and Muntenia. However, large areas have been reported in the sub- Carpathian hills of Moldova and in the hard permeable soils of Romanian.

- *Excess moisture caused by free level ground water*, fed from precipitation, infiltration from rivers, lakes, fish ponds, rice facilities, leaking underground and sometimes loss of irrigation water.

Water balance during periods of excess moisture relation is expressed as:

$$P + Inf + Ir - Et > S + I + W_{max}$$

where:

P - average annual rainfall;

P + Inf + Ir - Et > S + I + Wmax

Inf - infiltration from rivers, lakes and other sources; Ir - loss of water from irrigation systems;

ET - average annual evapotranspiration; S - surface water drainage of land;

I - the amount of water infiltrated in the soil profile, below the plant roots;

W_{max} - the maximum amount of water that a soil can accumulate in the plants root zone.

In this case, **the excess water volume** (**Ve**) is the appropriate minimum amount of porosity of aeration required for normal respiration of plant roots, respectively, of aerobic microorganisms.

Agricultural land with excess moisture from the category of soil water balance are commonly encountered in the irrigated or unirrigated territories from the Romanian Plain, more specific soils with good permeability (K> 0.5 m/day). These soils are formed on loess and sandy deposits where the water table is located at medium depths, less than 2-4 m.

- *Excess humidity caused by rain and ground water with high, slightly ascending level*, supplied from precipitation, infiltration from rivers, lakes, fish ponds, rice facilities, underground leaking and others, in low permeable clay soil conditions. Water balance during periods of excess moisture is obtained by the relation:

where:

P - average annual rainfall;

$$P + Af - Et > S + I + Wmax$$

Af - water intake from the phreatic layer under pressure; ET - average annual evapotranspiration;

S - surface water drainage of land;

I - the amount of water infiltrated in the soil profile, below the plant roots;

W_{max} - the maximum amount of water that soil can accumulate in the plants root zone.

The agricultural land with this type of moisture excess is common on low land in the Banat, Crisana and Maramures, in internal river meadows and some areas in the Danube.

After the estimated forecast by *National Meteorological Administration* in the coming years is likely to increase the size of the droughts that will occur in alternation with periods of intense rainfall. During relatively short periods of heavy rainfall, flooding will occur quickly in river meadows with different intensities inside and manifestation of excess moisture in the soil.

The effect of increasing average temperatures predicted for the coming years is considered to be more pronounced in areas of the Romanian Plain and less significant in sub- mountainous and mountainous areas of the Carpathian chain.

3. EXPERIMENTAL DRAINAGE SITE - BAIA DEPRESSION

Water excess in soil is a complex process that is determined by some factors like water uptake, circulation and removal in the system soil-subjacent rock. For removing water excess from soil profile, drainage systems were planned for improving soil aeration regime and field cultivation.

For knowing the long-term effect of hydro-ameliorative and soil-ameliorative works, applied under conditions of soil from drainage fields arranged in Baia, was determined the evolution of main eco-soil indicators of unimproved and improved soils.

3.1 Soil genetic conditions

The Baia Depression is located in the Sub-Carpathian Basin of the Moldova River, which is located in Suceava County, Romania. The total area of the Baia Depression is approximately 15,000 ha, 5,000 ha of which have excess soil moisture owing to rainfall or groundwater capillary rise.

The Baia Drain Field has an area of 3 ha and is located on a terrace platform of the Moldova River. The area bends slightly to the southeast and has a longitudinal slope of 2 - 5% and a height of 393 m. The parental material is silty clay with a thickness of 10 - 15 m. As a result, the natural drainage of the soil is weak to very weak. In addition, the groundwater is situated at a mean depth of 9 - 10 m, and the perched water table is located at a mean depth of 0.2 - 0.5 m.

The climate of the study area is continental temperate, with great thermal amplitude and rainfall that occurs primarily in the vegetative season. The rainfall in the study area is generally spread non-uniformly and usually torrential. The mean multiannual temperature from 1978 - 2008 was 7.9 ºC, the average rainfall was 806 mm and the average evapotranspiration was 599 mm, with an average water excess of 207 mm.

The vegetation in the study area was composed of a hygrophilous environment that contained an association of *Agrostis tenuis* and a sub-association of *Deschampsia caespitosa, Juncus effuzus*, and *Carex sp.*

Evaluation of the study location revealed that the area was characterized by albic stagnic - glossic Luvosol under the *Romanian System of Soil Taxonomy* (Florea et al., 2003) or albeluvisol under the *World Reference Base for Soil Resources* (WRB, 2006).

3.2 Tile drainage experimental variants

To optimize the air and water regime of soil in the area, an experimental field containing a tile drainage system designed to remove excess water was constructed in 1978.

During the study period (1978 - 2010), six different tiles drainage systems (A, B, C, D, E, F) were installed and the effect of the removal of excess moisture from the soil by these systems was evaluated (Figure 1).

Moreover, in variant A, the tile drainage was combined with the land forming in the bedding system with ridges and furrows. In this variant, furrows were constructed upon tile drains with spacing at 20 m. The bedding turning furrows to the middle of a cut to form a ridge that gradually slopes toward deep furrows to ensure good drainage.

Each drainage variant includes three drain lines. The length of each drain was 100 m and the slope of the drain lines was 0.2%.

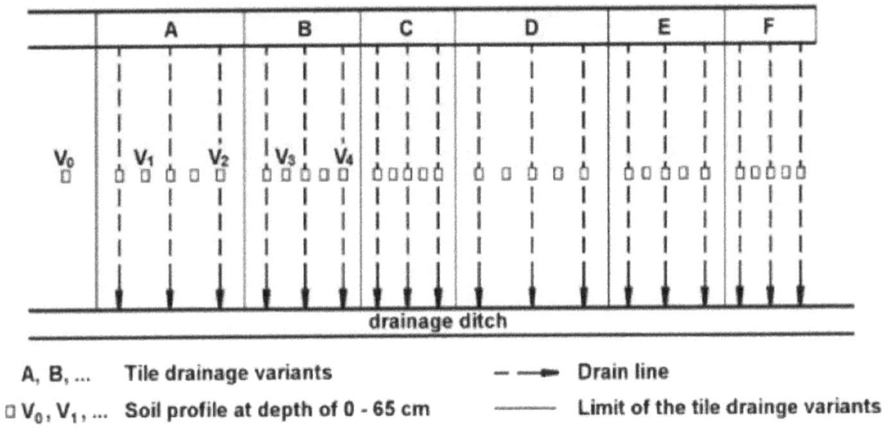

Fig. 1. Hydrotechnical scheme of the Baia drainage experimental site.

In variants A, B and C the depth of the tile drains is 1.0 m. The spacing between drains is 20 m in variant A, 15 m in B variant and 12 m in C variant. In variants D, E and F, the depth of the tile drains is 0.8 m. The spacing of the drains is 20 m in variant D, 15 m in variant E and 12 m in variant F.

Hydrotechnical scheme of the Baia drainage experimental site (Figure 1) was sized to optimize the basic design elements, used routinely in the design of tiles drainage systems (Table 1).

Table 1. Materials used for construction of the Baia drainage experimental site.

Tile drainage variant	Spacing between drain lines /depth drain (m)	Drain line number	Pipe type and diameter (mm)	Type and thickness of complex drain + filter (cm)
A	20/1,0	1 2 3	Tile Ø 70 Tile Ø 125 Tile Ø 70	Ballast (20) + Flax stems (50) Ballast (70) + Green sods Ballast (20) + Green sods
B	15/1,0	4 5 6	Tile Ø 70 Tile Ø 70 Tile Ø 70	Flax strains (30) Ballast (12) + Flax stems (20) Ballast (15) + Green sods
C	12/1,0	7 8 9	Corrugated plastic Ø 65 Smooth plastic Ø 63 Tile Ø 70	Ballast (12) + Green sods Ballast (12) + Green sods Ballast (15) + Green sods
D	20/0,8	10 11 12	Corrugated plastic Ø 65 Smooth plastic Ø 110 Tile Ø 70	Ballast (20) + Flax stems (40) Ballast (60) + Green sods Ballast (20) + Green sods

Tile drainage variant	Spacing between drain lines /depth drain (m)	Drain line number	Pipe type and diameter (mm)	Type and thickness of complex drain + filter (cm)
E	15/0,8	13 14 15	Tile Ø 70 Tile Ø 70 Tile Ø 70	Flax strains (30) Ballast (12) + Flax stems (20) Ballast (15) + Green sods
F	12/0,8	16 17 18	Corrugated plastic Ø 65 Smooth plastic Ø 63 Tile Ø 70	Ballast (12) + Green sods Ballast (12) + Green sods Ballast (15) + Green sods

- *Spacing between drain lines:* 20 m (variants A and D), 15 m (variants B and E) and 12 m (variants C and F) were determined assuming a non-permanent flow regime.
- In terms of functional efficiency, the spacing between drain lines, should allow lowering the groundwater level from the maximum height to optimum height, within
- 3-5 days. This lowering of groundwater level in time guaranteed certain plant root zone aeration.
- *The depth of the tile drains:* **1.0 m** (A, B and C variants), **0.8 m** (variants D, E and F) was sized taking into account the horizontal position of the hard permeable soil profile. If the soil is albic stagnic - glossic Luvosol, the Bt1W textural horizon is located at a depth of 0.9 to 1.0 m.
- *Pipe type and diameter* that was used for the construction of all 18 drain lines included the following materials: *tile drain* (pipe are from ceramic) with inner diameter of 70 mm and 125 mm, tube length of 33 cm with circular inner section and outer section of hexagonal form, *smooth plastic drain* with external diameter of 63 and 110 mm; *corrugated plastic drains* with linear waves and external diameter of 65 mm.
- *Filter material* that was put on top and around the drainage tube consisted of the following: river ballast, and, for some certain drain lines Ballast with flax strains and vegetal soil on the form of green sods.

Under Romania natural conditions the ballast from the river inland meadows is considered al widely used material as a filter in tube drainage works. Filter layer of Ballast has very good drainage effect because of the maintenance of its permeability over the period of operation of the drainage system. Ballast placed around and above the drainage pipes does not compact under soil conditions; it is not degraded by microorganisms in the soil or groundwater chemical action.

Implementation of drainage tubes was performed using E.T.T. 202-A equipment, a mechanic system that performed: digging trenches with a width of 50 cm, variable depth of 0.8 – 1.5 m and laying Tile drainage pipes and / or plastic.

After estimating the cost value of the lei / ha, on the time of implementation of drainage experimental variations has resulted some differentiate constructive solutions. For drain lines: 11/variant D, 2/variant A was the highest assessed values of the cost. Lower cost price values were obtained for the drains: 13/variant E and 12/variant D, depending on dimensional elements and building materials used.

3.3 Improvement procedures

After installation of the tile drainage system, the study area was used for farming.

The drained soil was then improved by applying the three following experimental cycles:

- *Cycle 1* (1978-1986): superficial land forming, cultivation of virgin hygrophilous meadow, deep loosening to a depth of 0 - 70 cm, amendment by the application of 10 - 12 t·ha-1 limestone with a content of 95 - 100% CaCO , organic fertilizing with 40 t·ha-1 manure (in variants A, B, C, D, E and F), land forming in the bedding system with a ridge and furrow (only in variant A) once a cycle and annual mineral fertilization (in all variants);
- *Cycle 2* (1987 - 1997): deep re-loosening at the useful depth of 0 - 70 cm, re-amendment of the soil with 7 - 8 t·ha-1 limestone, organic refertilization with 40 t·ha-1 manure once a cycle and annual mineral fertilization;
- *Cycle 3* (1998 - 2010): no application of soil improvement and/or cropping works.
- Crop rotations and plants were cultivated in the following three experimental cycles:
- *Cycle 1*, crop rotation I: maize-two-row barley-potato-flax fibers (1979 - 1982); crop rotation II: (wheat + rye), (maize + potato), maize - potato (1983 - 1986);
- *Cycle 2*, mixture of seeded perennial grasses (70%) and legumes (30%) (1987 - 1990); mixture of perennial grasses (70%) and overseeded legumes (30%) (1990 - 1997);
- *Cycle 3*, overseeded natural meadow (1997 - 2010).

To quantify the changes in the major physical and chemical characteristics of the improved soil in response to the following variations of the drainage treatments described above (Figure 1), at the end of each cycle soil samples from the following treatments were analysed:

V_0 - control variant, natural unimproved meadow (Figure 2);

V_1 - tile drainage combined with land forming in the bedding system variant: top of ridges (Figure 3);

V_2 - tile drainage combined with land forming in the bedding system variant: drain line - furrow cross-section (Figure 3);

V3 - tile drainage variant: middle of the drain lines (Figure 4); V4 - tile drainage variant: drain line cross-section (Figure 4).

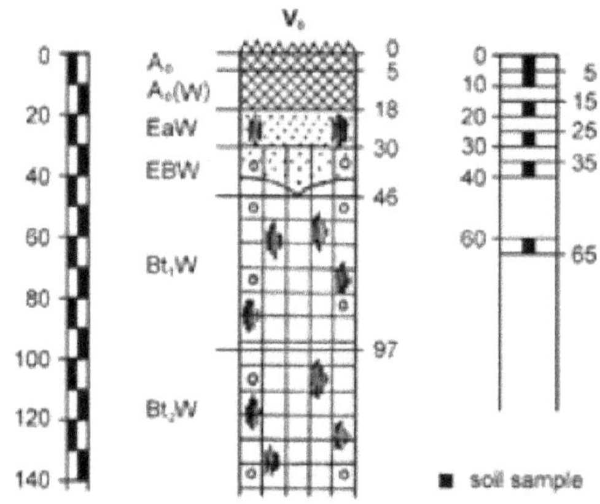

Fig. 2. Soil sampling in the control variant (V0).

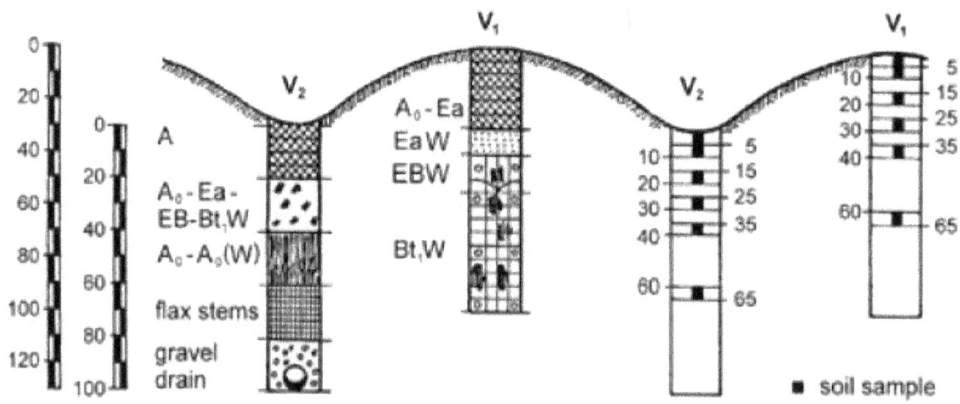

Fig. 3. Tile drainage combined with the bedding system variant - top of ridges (V1) / furrow (V2).

The evolution of some physical and chemical features under the influence of the improvements was also analyzed in the three experimental cycles. To accomplish this, soil samples were collected from the ploughed stratum (0 - 20 cm) and the horizon below this layer (20 - 40 cm). Soil samples were collected using an agrochemical drill at the end of each experimental cycle in 1986, 1997 and 2010 from the entire surface of the experimental variants (V0, V1 and V3).

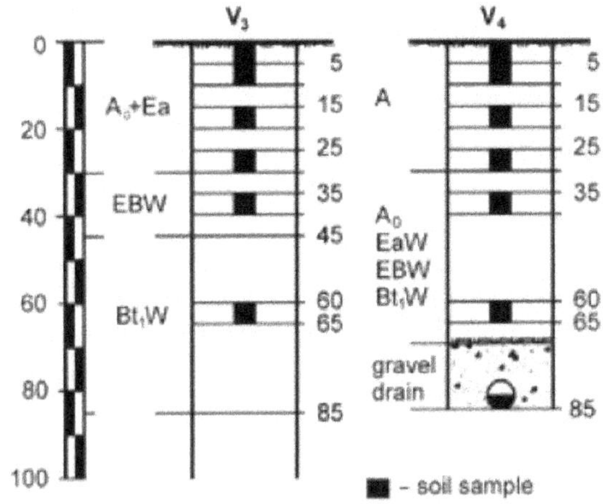

Fig. 4. Tile drainage variant - middle of the drain spacing (V3) and drain cross-section (V4).

In the natural conditions of this wetland, the albic stagnic - glossic Luvosol, which is used as a natural pasture, presented the following morphological features in the horizon succession (Figure 2):

A0 0 - 5 cm - strongly unreclaimed stratum; silt loam; very dark brown grey 10 YR 4/2 (wet) and light grey 10 YR 7/2 (dry); small polyhedral angular structure, moderately developed; very friable in wet state, quite hard in dry state; weakly plastic; weakly adherent; very frequent roots.

A0 (W) 5 - 18 cm - silt loam; dark greyish brown 10YR 4/2 (wet) and light greyish brown 10YR 6/2 (dry); small polyhedral angular structure, moderately developed, friable in wet state and easily compacted when dry; weakly plastic; weakly adherent; frequent coarse and medium pores; thin and very thin frequent roots; clear contact with the horizon below.

EaW 18 - 30 cm - silt loam; light olive colour 7.5Y 6/2 with small yellowish brown spots 10 YR 6/6 (wet) and white 7.5 Y 8/2 with yellowish brown mottles 10 YR 6/8 (dry); small polyhedral angular up to lamellar structure, weakly developed, friable in wet state and easily compacted when dry; weakly plastic; weakly adherent; frequent medium pores; rare very thin roots; diffuse irregular contact with the horizon below.

E+BW 30 - 46 cm - clayey loam; greenish gray 5GY 6/1 with small and medium frequent mottles, light yellowish brown 10 YR 6/4 (dry) and white 7.5 Y 8/1 with brown yellowish spots 10 YR 6/6 (dry); medium polyhedral angular structure, moderately developed; clay films on the faces of structural aggregates; small and medium ferromanganese spots and concretions; tough when wet and compacted in dry state; plastic; adherent; frequent pores; diffuse irregular contact with the horizon below.

Bt1W 46 - 97 cm - loamy clay, greenish gray 5GY 6/1 with frequent large yellowish brown mottles 10 YR 5/6 (wet) and light gray 7.5 Y 7/1 with light yellowish brown 10 YR 6/4 and brown-yellowish 10 YR 6/6 mottles (dry); large prismatic structure, moderately developed; coarse clay films on the faces of the structural aggregates; medium ferromanganese spots and concretions; hard in wet state and very compacted when dry; plastic; adherent.

Following placement of the tile drainage system, land was forming in the bedding system and then the soil was cultivated for the first time. In consequence, a mixture between the upper genetic horizons of the soil profile was observed. After the first experimental cycle (1978-1986), the soil from the surface formed in the bedding system presented the following horizon succession (Figure 2 and Figure 3):

A0 + **A0 (W)** + **EaW** - (0-30 cm): silt loam;

EaW - 30-40 cm: silt loam;

Ea+BW - 40 - 55 cm: clayey loam;

Bt1W - 55 - 97cm: loamy clay.

In the new conditions of the improved soil, the active physiological depth was greater than that in the unimproved soil. The succession of the horizon morphology revealed that the Bt1W textural horizon is situated at a higher depth when compared to the unimproved soil.

The experimental parcels from the space between the drain lines also showed a mixture of the soil profile upper horizons. The changes inferred were signalled in the depth of the ploughed horizon (0 - 30 cm), and occurred due to the agricultural cultivation works and the deep loosening from the first two experimental cycles (1978 - 1986 and 1987 - 1996). The space between the tile drain lines resulted in the following horizon succession (Figure 4):

(A0 + Ea)p - 0 - 30 cm: silt loam; **Ea+BW** - 30 - 46 cm: clayey loam; **Bt1W** - 46-97 cm: loam clay.

The analysed physical and chemical characteristics of soil sampled in 2010 were determined according to the norms of the *Methodology of soil studies elaboration - National Institute of Research and Development in Soil Science, Agrochemistry and Environment Bucharest* - 1987 and to the *Romanian System of Soil Taxonomy* - 2003.

3.4 Hydrometeorological balance of field drainage pipes

Depending on **annual precipitation (P)** and **annual water consumption by evapotranspiration (ET)** was expressed, on the basis of made observations during the experiment, the annual hydrometeorological balance (**± ΔP = P - ET**).

- *Annual rainfall regime* (P) was the main source of excess water from the ground, being based on alternation between periods of intense rainfall and, respectively, dry periods in May. Database of observations made on annual rainfall were extracted for the case study examples, from a rainy year (1981), respectively, a dry year (1982).

For wet weather conditions area of north-eastern Romania, was considered as rainy year, 1981, with an annual rainfall of 968 mm fallen (statistical insurance 5%) and the dry year, 1982, with annual rainfall of 742 mm insurance and 80% (Table 2)

The average volume distribution of 855 mm from the years 1981-1982 was 541 mm (63 %) in summer and 314 mm (27 %) in winter.

Table 2. Annual rainfall and seasonal distribution in 1981-1982.

Year	Annual rainfall (P mm)	Season rainfall distribution (mm)	
		Warm season months (V-X)	Cold season months (XI-IV)
1981	968	592	376
1982	742	489	253
Average	855	541	314

- *Average multi-annual rainfall* throughout the period examined in the years 1978-2010 recorded an annual average value of 806 mm (50% assurance).

In terms of annual rainfall distribution and intensity of growing seasons, months and calendar time intervals from one day up to 5 consecutive days of the years 1978-2010 have been reported following characteristics:

- In the hot season (V-X) the annual average year, rainfall was recorded 549 mm (68%), and in winter (XI-IV), 257 mm (32%).
- Average and maximum monthly rainfall recorded multi-annual, usually, recorded the largest amounts of water in June and July and lowest in January-March.
- Rainfall in periods of 1-5 consecutive days were characterized, throughout the study period, by the size and distribution on everyday, 2 days, 3 days and 5 days, depending on the overall progress of the climatic conditions (Table 3).

Table 3. Multiannual precipitation (1978-2010), average and maximum, monthly and on intervals of 1-5 consecutive days.

Month of the year	Monthly precipitation (mm)		Rainfall in consecutive intervals of 1-5 days (mm)							
			1 day		2 days		3 days		5 days	
	average	max.	average	max.	average	max.	average	max.	average	max.
I	33	69.3	12	25.7	16	31.3	18	36.5	20	49.2
II	34	80.2	11	18.9	14	23.2	16	30.0	18	33.5
III	33	63.0	11	25.1	12	25.1	14	25.1	16	28.3
IV	80	12.3	26	41.3	33	53.6	38	65.5	45	79.9
V	97	199.1	32	95.1	43	95.1	46	97.5	50	97.5
VI	133	193.2	49	72.0	65	96.5	78	140.8	89	153.8
VII	133	236.5	40	89.1	51	100.7	54	109.6	62	137.0
VIII	82	186.2	32	43.6	39	81.8	45	118.5	53	158.7
IX	63	125.8	21	37.5	28	59.8	30	59.8	35	66.5
X	41	127.5	16	39.5	21	73.4	21	75.8	24	98.2
XI	41	105.1	18	38.5	23	59.4	24	61.4	26	62.2
XII	36	103.8	12	25.6	16	39.9	17	41.0	20	43.7
Annual	806	-	280	-	361	-	401	-	458	-

- *Annual evapotranspiration regime* (*ET*), ranged between the limits of normal distribution of the wet weather conditions. For the two characteristic years of the climate conditions (1981 and 1982), annual evapotranspiration was 559 mm (1981) and 580 mm (1982), with an average of 569 mm. Distribution of mean annual evapotranspiration for the two growing seasons of the year difference between the value of 517 mm (91%) in the warm season, respectively, 52 mm (9%) in cold season of the year (Table 4).

Table 4. Annual and seasonal evapotranspiration in years 1981 - 1982.

Year	Annual evapotranspiration (Et mm)	Seasonal distribution of evapotranspiration (mm)	
		Warm season month (V-X)	Cold season month (XI-IV)
1981	559	505	54
1982	580	529	51
Average	569	517	52

- *Average annual evapotranspiration regime* throughout the period examined in the years 1978-2010 recorded an average of 599 mm per year which varied in size according to the general trend of climatic factors and the use of agricultural land.
- Distribution of average annual evapotranspiration (1978-2010) on the two growing seasons, differentiate significantly separately. In the warm season (V-X), which corresponded to wet land with vegetation period of crops, has developed a water consumption of 481 mm, which means 80% of annual average. In winter (XI-IV), evapotranspiration recorded in terms of annual average, water consumption of 118 mm, which is 20%.
- Average monthly evapotranspiration was characterized by maximum values during the months of summer season with a 90-142 mm consumption, small to medium in the months of spring - autumn, with a consumption of 1-76 mm and negligible values months of winter.
- Evapotranspiration and diurnal periods of up to 3-5 consecutive days, was typically, below the rainfalls in these intervals, which resulted in a surplus in the hydrometeorological water balance.
- The amount of excess water with other natural factors contributed to the formation and maintenance of the soil profile or on the land of excess rainfall from moisture.
- *Annual hydrometeorological regime balance* ($\pm \Delta P = P - Et$) registered in terms of the years 1981 to 1982 an average surplus of 286 mm, 162 mm annual distribution in 1982 and 409 mm in 1981 (Table 5) .

Distribution of surplus and/or water deficit equation from annual hydrometeorological balance individualized only by excess water in 1981, respectively by excess and water deficit in 1982. The distribution of the two seasons average surplus of 286 mm to 262 mm difference between the value (92%) in winter and 24 mm (8%) in summer.

Table 5. Annual hydro-meteorological balance and seasonal distribution in 1981-1982.

Year	Annual hydro-meteorological balance				Warm season months (V-X)				Cold season months (XI-IV)			
	P (mm)	Et (mm)	± ΔP (mm)		P (mm)	Et (mm)	± ΔP (mm)		P (mm)	Et (mm)	± ΔP (mm)	
			+	-			+	-			+	-
1981	968	559	409	-	592	505	87	-	376	54	322	-
1982	742	580	162	-	489	529	-	40	253	51	202	-
Average	855	569	286	-	541	517	44	20	314	52	262	-

- *Hydrometeorological annual average regime balance* of the period under review (1978-2010) recorded a surplus of 207 mm water, characterized by the following features on the seasons, months and characteristic times of the year.
- Distribution of the two growing seasons of the year differenced between 139 mm (67%), in the hot season of the year annual average and 68 mm (33%) in winter.
- Annual distribution of water surplus resulted from the hydrometeorological balance was characterized in terms of environment, positive in almost all months of the year.
- The average values of the determined period were employed between 0-11 mm minimum quantities, made in July, September and October and 31-40 mm maximum on the other months of the year, except August.
- In terms of climate, we can say that the August register, usually an average deficit of 11- 20 mm, offset by reserves of water stored in soil in winter. In this context, water scarcity may be an example of 40 mm was recorded in August of 1982, considered a dry year.

4. TECHNICAL EFFICIENCY OF THE EXPERIMENTAL DRAINAGE VARIANTS

4.1 Periods of drainage system operation

In periods of drainage operation with non-permanent flow regime, rainfalls volume (P) exceeded the water consumption by evapotranspiration (Et), deep infiltration (I) and drain the land surface (S). In periods of excess water removal (ΔP), drained soil is working functionally as a reservoir, which records under the effect of heavy rains filling, respectively, depletion under the effect of drainage.

To characterize evolution of climatic factors that determine the hydro-meteorological balance and proper functioning of pipe drainage are presented the observations from the operation of drainage tubes in the years 1981-1982 (Table 6).

Hydrometeorological balance, that helped triggering the operation of the drainage system with pipes, had on given weather conditions between 1981-1982, the following significant issues:

- Excess water (ΔP) derived from annual hydrometeorological balance recorded 625 mm in 1981 and 401 mm in 1982. This water surplus caused the total filling of the lacunar space of soil, which led to its discharge in the drainage tubes system.
- Dynamics of water surplus recorded in intervals of drainage operation was located, under review conditions, between the minimum values of 1-4 mm and maximum of 102-113 mm.
- The functioning duration of drainage pipes of periods with water surplus totaled from 2 days to 20 consecutive days, as determined by the volume, duration and intensity of rainfalls.

4.2 Specific drainage flows

To characterize the dynamics of drainage specific flows, first were analyzed the observations from climatic conditions in 1981 and 1982. In periods of operation of drainage mentioned in the previous paragraph, specific drainage flows (q) expressed in l/s and ha for each drain line were characterized, generally, depending on specific factors.

Specific flow drainage dynamics was analyzed depending on the size of the excess rainwater and drainage lines efficiency, respectively, on the average assessed for the six experimental drainage variants (A, B, C, D, E and F). Also, tube drainage operation was highlighted both in terms of association with land shaping in the bedding system with ridges and furrows (variant A) and, respectively, aeration soil to a depth of 50-60 cm on average (variants B, C, D, E and F). In the early hours of operating periods of pipes drains were measured, typically, maximum flow, which measure on the evolution of precipitation increases and/or continuity of the excess volume of water till minimum values.

Table 6. Operating periods of drainage tubes in the years 1981 - 1982.

Year	Drainage period	Number of days	Hydrometeorological balance		
			P (mm)	Et (mm)	+ΔP (mm)
1981	3-17 II	14	70	1	69
	9-29 III	20	100	10	90
	27-30 IV	3	22	7	15
	6-18 V	12	135	22	113
	21-24 V	3	18	6	12
	10-19 VII	9	90	38	52
	21-25 VII	4	22	18	4
	27-31 VII	4	53	16	37
	1-4 VIII	4	47	18	29
	16-19 IX	3	60	6	54
	21-30 XI	9	105	3	102
	7-15 XII	8	20	-	20
	16-31 XII	15	28	-	28
	Annual	108	770	145	625
1982	12-31 III	20	122	10	112
	1-5 IV	4	3	2	1
	19-24 IV	5	20	5	15
	28-30 IV	2	40	3	37
	1-5 V	5	20	4	16
	15-22 VI	7	108	25	83
	25-29 VI	4	22	14	8
	1-3 VII	2	24	8	16
	11-22 VII	11	95	46	49
	23-25 VII	2	21	10	11
	26-31 VII	6	56	21	35
	1-10 VIII	9	50	32	18
	Annual	77	581	180	401

Specific flow regime of pipe drains drainage tubes was characterized on the base of volumetric flows measured 2-3 times daily, on the discharge place of each drain, in drainage operation during the years 1981 and 1982 (Table 7).

Table 7. The specific drainage flows in 1981-1982.

Variant and spacing between drain lines /depth drain (m)	Drain line number	Specific drainage flow q sá(l^{-1}haá$^{-1}$)								
		Maximum extreme flows			Annual maximum average flows			Annual average flows		
		1981	1982	Average	1981	1982	Average	1981	1982	Average
A	1	5.17	4.87	5.02	1.89	1.25	1.57	0.63	0.31	0.47
	2	5.77	5.17	5.47	1.55	1.37	1.46	0.49	0.32	0.41
20/1.0	3	5.00	4.84	4.92	1.46	1.21	1.34	0.48	0.30	0.39
	Average	5.31	4.96	5.14	1.63	1.28	1.46	0.53	0.31	0.42
B	4	2.98	4.44	3.71	0.95	1.27	1.11	0.36	0.33	0.35
	5	2.90	3.92	3.41	0.99	1.25	1.12	0.36	0.32	0.34
15/1.0	6	2.86	4.26	3.56	1.02	1.30	1.16	0.36	0.32	0.34
	Average	2.91	4.21	3.56	0.99	1.27	1.13	0.36	0.32	0.34
C	7	3.85	6.25	5.05	1.30	1.80	1.55	0.48	0.41	0.45
	8	3.97	5.95	4.96	1.39	1.74	1.57	0.48	0.42	0.45
12/1.0	9	4.17	6.58	5.38	1.38	1.66	1.52	0.48	0.46	0.47
	Average	4.00	6.26	5.13	1.36	1.73	1.55	0.48	0.43	0.46
D	10	2.50	4.17	3.34	0.87	1.24	1.06	0.31	0.29	0.30
	11	3.13	3.49	3.31	0.95	1.15	1.05	0.33	0.28	0.31
20/0.8	12	2.94	3.20	3.07	0.96	1.01	0.99	0.33	0.27	0.30
	Average	2.86	3.62	3.24	0.93	1.13	1.03	0.32	0.28	0.30
E	13	4.00	5.41	4.71	1.23	1.69	1.46	0.44	0.41	0.43
	14	3.33	5.71	4.52	1.22	1.51	1.37	0.42	0.40	0.41

Variant and spacing between drain lines /depth drain (m)	Drain line number	Specific drainage flow q sá(l^{-1}haá$^{-1}$)								
		Maximum extreme flows			Annual maximum average flows			Annual average flows		
		1981	1982	Average	1981	1982	Average	1981	1982	Average
15/0.8	15	4.00	5.13	4.57	1.32	1.59	1.46	0.45	0.37	0.41
	Average	3.78	5.42	4.60	1.26	1.60	1.43	0.44	0.39	0.42
F	16	4.55	6.10	5.33	1.48	1.80	1.64	0.54	0.43	0.49
	17	4.55	6.76	5.66	1.61	1.93	1.77	0.60	0.46	0.53
12/0.8	18	4.17	6.07	5.12	1.31	1.73	1.52	0.53	0.40	0.47
	Average	4.42	6.31	5.37	1.47	1.82	1.64	0.56	0.43	0.50

- *The maximum extreme flows* sá, measured in the first functioning hours of drainage under the conditions of the 18 drains lines showed obvious differences. In terms of the year 1981, measured flows were hierarchized between 2.50 l-1 sáand ha (drain 10/variant D) and 5.77 l-1 sáand ha (drain 2/variant A), and in 1982 from 3.20 l-1 sáand ha (drain 12/variant D) and 6.76 l-1 and ha (drain 17/variant F).
- *Annual maximum average flows* sáwere individualized on the 18 line pipe drains between the minimum value of 0.87 l-1 sáha (drain 10/variant D) and a maximum of 1.89 lá- 1 sáha (drain 1/variant A), during operation of drainage in 1981. There is a similar dynamic in terms of 1982, with a variation of this flow from 1.01 lá-1 sáha (drain 12/variant D) and up to 1.93 lá-1ha (drain 17/variant F).
- *Annual average flows,* that were expressed on the basis of all measurements from the 108 days in 1981 and 77 days in 1982 showed also, functional technical efficiency of solutions designed in the 18 lines of drainage.

Mean annual specific drainage flows were correlated with the size of maximum flow and maximum annual extreme, both for individual conditions of drains, and also for the estimated average experimental version of drainage lines consisted of three drains lines. Thus, in 1981 the annual average flows ranged from 0.31-1ha (drain 10/variant D) and sá0.63 -1 sáha (drain 1/variant A) and in 1982, from 0.27á-1 sáha (drain 12/variant D) and 0.46á-1ha (drain 17/variant F).

Technical efficiency of drainage variants, depending on the size of the specific flows highlighted the following issues relevant to the experimental cycle:

- The size of discharged flows showed a higher efficiency of drains to the distance between them 12 m and average depth of 0.8 m and 1.0 m (F and C variants), who drained a 20% to 70% more, compared to variants with distance between drains of 15 m and 20 m.
- Association of the pipe drainage with the soil shaping combined with the bedding system at the same distance of 20 m (variant A) provided a capture and efficient evacuation of excess water, comparable to the distance between drains variant F 12 m and average depth of 0.8 m.
- The usage of a filter layer as a prism with dimensions of 20 x 70 cm was achieved from Ballast and flax strains on drained 1/A, and from ballast on drain 2/A, ballast layer with a thickness of 20 cm, on drained 3/A did not show a high functional efficiency.
- The increase of the distance between the drains or drain depth settlement resulted in a reduction of debts discharged, except association of pipe drainage tubes with land shaping in the bedding system with ridges and furrows (variant A).

4.3 Drained water volumes

Depending on the size of specific drainage daily flow of the area served by the drainage lines (1-18) and experimental drainage variants (A-F) volumes of drained water were calculated, expressed in mm or m3haá-1. To calculate the volume of drained water areas were used areas of 0.60 ha (variants A and D), 0.45 ha (variants B and E), and 0.36 ha (variants C and F).

Average annual volumes of drained water (Table 8), in the excess water conditions during 1981 and 1982 showed a differentiated technical efficiency, first drain lines respectively arranged on the six variants, as follows:

- The individual efficiency of the component drains of the network drainage was illustrated by the size of drained water volumes, which ranged from 232 mm (2320 m3/ha) for drain 12 / D and maximum of 406 mm (4 060 m3/ ha) for drain 1 / A, associated with land shaping in the bedding system with ridges and furrows.
- Operational efficiency of the experimental variants, depending on the size of individual volumes of component drains was noted by a reduced capacity for variant D (20/0.8 m) with a drain rate of 245 mm (2450 m3haá-1) and better under variant F (12/0.8 m), which achieved a 368 mm drained rate (3680 m3haá-1).
- *Technical efficiency of drainage variants*, depending on the size of the drained water volumes was materialized through the following significant features:
- Operating mode of the drains lines is dependent on seasonal rainfall distribution and consumption of water by evapotranspiration.
- Maintenance of the same spacing between the drain lines of 12 m or 15 m and increase of the depth of drains from 0.8 m to 1.0 m contribute in reducing the excess capacity of water interception by an average of 3% for a distance of 12 m (variant F to variant C) and 22 % for distance of 15 m (variant E to variant B).
- Drained soil aeration to an average depth of 50-60 cm resulted in an improvement in rainfall interception excess water by about 2-9%, according to the first two lines of drains (B, C, D, E, F variants) than the unaerated surface soil of the third drain.
- Average volumes of drained water from the corresponding size of the excess moisture from rainfall from 1981 and 1982, on the 6 variant were distributed among the 48% efficiency (variant D) and up to 72% (variant F).
- Drained water volumes were found in large quantities during the cold season (XI -IV) and lower in the warm season of the year (V -X).

Table 8. Drained water volumes regime in 1981-1982.

Spacing between drain lines /depth drain (m)	Drain line number	The annual volume of drained water (mm)			Seasonal distribution of drained water volumes (mm)					
					Warm season (V-X)			Cold season (XI-IV)		
		1981	1982	Average	1981	1982	Average	1981	1982	Average
A 20/1,0	1	583	229	406	202	102	152	381	127	254
	2	469	211	340	185	110	147	284	101	192
	3	395	178	287	166	82	124	229	96	163
	Average	482	206	344	184	98	141	298	108	203
B 15/1,0	4	359	207	283	143	114	128	216	93	155
	5	360	205	283	142	111	126	218	94	156
	6	352	192	272	144	98	121	208	94	151
	Average	357	201	279	143	107	125	214	94	154
C 12/1,0	7	456	273	364	177	142	159	279	131	205
	8	453	273	363	175	139	157	278	134	206
	9	433	248	341	164	117	140	269	131	200
	Average	447	265	356	172	133	152	275	132	204
D 20/0,8	10	310	189	249	125	100	112	185	89	137
	11	327	183	255	128	100	114	199	83	141
	12	324	139	232	134	69	101	190	70	130
	Average	320	170	245	129	90	109	191	80	136

Spacing between drain lines /depth drain (m)	Drain line number	The annual volume of drained water (mm)			Seasonal distribution of drained water volumes (mm)					
					Warm season (V-X)			Cold season (XI-IV)		
		1981	1982	Average	1981	1982	Average	1981	1982	Average
E 15/0,8	13	432	258	345	171	137	154	261	121	191
	14	440	240	340	177	118	148	263	122	193
	15	430	239	335	169	122	145	261	117	189
	Average	434	246	340	172	126	149	262	120	191
F 12/0,8	16	457	279	368	182	148	165	275	131	203
	17	482	296	389	196	155	176	286	141	214
	18	428	265	347	171	136	154	257	129	193
	Average	456	280	368	183	146	165	273	134	203

- The exploitation behaviour of the drainage lines of the experimental variants, showed an increase in water drained in variants when the distance between drains variants was 12 m and average depth of 0.8 m or 1.0 m.

4.5 The capture capacity of drain pipes

The excess stagnant water from the surface of the field and/or as pedophreatic water in the upper part of the soil structure must be evacuated in a relatively optimal period of time.

Before the descendent water current from the soil structure gets into the drain pipe it encounters a series of resistances that in general depend of the following natural factors and arrangement conditions:

- The permeability of the soil arranged with pipe drainage works;
- The permeability of the soil in the cross-sections of drain lines;
- The type and thickness of complex drain + filter;
- The pipe type and diameter;
- The distribution and the surface of the joints / perforations that allow the water to enter into the drain pipes.

According to *van Someren (1964),* the total energy losses in pipes (Δh) that are due to the resistances that appear when the water enters the drain pipe and that is measured at the middle of the distance between the drains, can be divided into the following three categories.

- *Horizontal energy losses (Δh0)* that are to the resistances encountered by the water current at the passage through the soil, up to approximately 1.0 m away from the drain pipe.
- *Radial energy losses (Δhr)* that is due to the convergence of the water current lines, from the immediate area of the drain pipe.
- *Entrance energy losses (Δhi)* that are determined by the resistances encountered by the water current when entering into the drain pipe.

For the water current to enter into the drain pipe, the total energy losses *(Δh)* from the middle of the spacing between the drain lines must be higher than the sum of the partial energy losses:

$$\Delta h > \Delta h_0 + \Delta h_r + \Delta h_i$$

The studies made on the *total energy losses (Δh),* under the conditions of a non-permanent regime were analyzed in the hypothesis of knowing the period of time (t days) in which the water level must be lowered from the maximum height (h0) to the minimum height (h_t).

The effect of the total energy losses ($\Delta_h = h_0 - h_t$) was followed in the piezometric pipe situated at the middle of the spacing between the drain pipes, with the distance between them of 12 m and the medium depth of 0.8 m. At different periods of time (t), the total energy loss was represented, in general, by the decrease of the excess water level at the maximum height of the depression curve of 60 - 70 cm, the best height of 15 - 20 cm. At the same periods of time (t) the drained flows decreased from 9 - 10 l/minute to 0.1 l/minute.

- *The factor of the draining intensity* ($a = \frac{1}{J}$) was determined in the base of the measured

levels in the piezometer from the middle of the spacing between the drain lines, at the beginning and the end of the draining's functioning periods, with the help of the following relation:

$$a = \frac{1}{J} = \frac{2.3\left(\log h_0 - \log h_t\right)}{t}$$

where:

a - factor of the draining intensity (days-1)

h_0 - maximum height of the pedophreatic water level *ht* - minimum height of the pedophreatic water level t - period of time (days).

Depending of the amount of rain that fell, that in periods of 3 - 5 consecutive days, registered values of up to 60 - 140 mm, there were differentiated as well the values of the intensity draining factors. So, depending of the *total energy losses* (Δh), after the previously mentioned rains, the values of the draining intensity were of 4 - 6 days, during the hot season, and of 6 - 14 days, during the cold season (Table 9).

Table 9. The draining intensity factor depending of the total energy loss.

Time (t)	The height of the depression curve at the middle of the spacing between the drain pipes		Total energy losses	Factor $a = \frac{1}{J}$	Coefficient $J = \frac{1}{a}$
days	h0 (mm)	ht (mm)	Δh (mm)	days -1	-
5	690	300	390	0.167	6.0
7	660	200	460	0.171	5.8
5	620	290	330	0.152	6.6
4	620	270	350	0.208	4.8
5	620	190	430	0.236	4.2
11	520	240	280	0.070	14.3
2	380	250	130	0.209	4.8
4	320	200	120	0.117	8.5

- *The filtration coefficient (K)* dayáwas characterized by a very good permeability, in the arable layer of 0 - 25 cm (K = 3.2 m-1dayá), fact that assured the high degree of efficiency of the process of draining the excess water. In the case of the underlying horizon that was analyzed for the depth of 25 - 65 cm, the permeability was lower (k = 0.2 m-1), fact that determined the execution and the periodical renewal of the soil loosening process until the necessary depth of 70 - 80 cm.
- *The draining porosity* was correlated with the values that resulted in the case of the filtration coefficient, being characterized by medium to high values of 44 - 54 % of the arable layer (0 - 25 cm) and by small values of 2 - 6 % in the underlying layer of 25 - 65 cm.

The drainage porosity consists of pores that are larger than 10-30 µ. These pores are often filled with air and serve as a pathway for the excess water to be drained through infiltration. The 30 µ diameter is present in sandy soils, while the 10 µ diameter is found in medium and coarse grained soils. Utile porosity refers to the presence of pores that have a diameter ranging from 0.2 to 10 to 30 µ. These holes are capable of retaining either air or mobile water that can be accessed by plants. Inactive porosity refers to the presence of tiny holes, with diameters less than 0.2 µ, that trap water that cannot be accessed by plants (Canarache, 1990). Drained porosity, useful porosity, and inactive porosity are the fundamental constituents of total porosity. The soil compaction condition is characterized by the fluctuation of these three components, which are determined by specific circumstances. Organic and organo-mineral soils require a drained porosity larger than 20% in order to have favorable circumstances. However, achieving this requires the installation of a drainage system (Canarache, 1991).

4.6 The behaviour during the use of draining pipes

The drainage system in soils with high rainfall is determined by the size and random distribution of the amount of rainwater. During a sequence of consecutive time intervals, an accumulation of excessive water occurs in the soil and/or on its surface, which needs to be removed by a drainage system.

The evaluation of the water survey in the conditions of the existent *18 draining lines* (1, 2, 3,...., 18) and respectively, of the *six experimental drainage variants* (A, B, C, D, E and F) was analyzed taking into consideration the long time observations that were realized on the following factors:

- The daily precipitations;
- The average daily temperature of the air;
- The real maximum daily evapotranspiration;

- The daily water flow evacuated by drain pipes;
- The water volumes evacuated by drainage network.

The annual dynamic of the climatic factors from the period 1978 - 2010 that was also presented in the previous paragraphs respected the characteristic cs of the areas from the wet climate area of Romania, area that includes the Baia Depression as well.

The synthesis of the results obtained during the entire period of time when pipes drain experiments were made, can be characterized according to the annual and the seasonal dynamic of the climatic factors and the functional efficiency of the drainage system, as follows:

- *The multiannual precipitation regime* - it was characterized by a medium quantity of 806 mm, with an annual distribution between the maximum quantity of 968 mm (1981), with a 5 % assurance, and the minimum quantity of 455 mm (1986), with a 95% assurance.
- *The real maximum evapotranspiration,* having a medium multiannual value of 599 mm situated between the limits of the normal distribution from the wet areas.
- *Effective water surplus* that resulted from the equation of the water survey from the drained soil, from the periods of time the drainage system with pipes worked registered a medium multiannual value of 325 mm or 3250 m3haá-1 (Table 10).
- *The medium volumes of drained water* differentiated, in general, according to the dimensional elements and the nature of the construction materials used at the draining system. In the conditions of the equipping the 18 draining lines, the maximum norm of drained water 2070 m3haá-1 (64 % of the medium water excess) was reached at the drain 17/variant F. The minimum norm of 1110 m3haá-1 was found at drain 12/variant D with the distance between the drains of 20 m and the medium depth of the pipes of 0.80 m (Table 10).
- *The medium daily water flow* was individualized between the maximum values of up to day á3.5 mm-1 (drain 17/variant F) and the minimum values of 1.8 mm day-1 (drain 12/variant D), according to the materials that were used and the dimensional elements of the drainage system (Table 10).

Table 10. The regime of the volumes of drained water in the period 1978 - 2010.

Variant and spacing between drain lines / depth drain (m)	Drain line number	Pipe type and diameter (mm)	Surface of the pipes joints / perforation s (cm^{2}má-1)	Drained soil water balance			
				Mean water excess (m^{3}haá-1)	Mean drained water m^{3}haá-1	%	dayáMean daily water flow (mm-1)
A 20/1,0	1	Tile Ø 70	10-15	3250	1830	56	3.0
	2	Tile Ø 125	13-20		1610	50	2.7
	3	Tile Ø 70	10-15		1390	43	2.3
	Average		-	-	1610	50	2.7
B 15/1,0	4	Tile Ø 70	10-15	3250	1370	42	2.3
	5	Tile Ø 70	10-15		1460	45	2.4
	6	Tile Ø 70	10-15		1440	44	2.4
	Average		-	-	1423	44	2.4
C 12/1,0	7	Corrugated plastic Ø 65	15-30	3250	1870	58	3.1
	8	Smooth plastic Ø 63	15-30		1850	57	3.1
	9	Tile Ø 70	10-15		1770	55	2.9
	Average		-	-	1830	57	3.0
D 20/0,8	10	Corrugated plastic Ø 65	15-30	3250	1260	39	2.1
	11	Smooth plastic Ø 110	30-40		1360	42	2.3
	12	Tile Ø 70	10-15		1110	34	1.8
	Average		-	-	1243	38	2.1

Variant and spacing between drain lines / depth drain (m)	Drain line number	Pipe type and diameter (mm)	Surface of the pipes joints / perforation s ($cm^2 má^{-1}$)	Drained soil water balance			
				Mean water excess ($m^3 haá^{-1}$)	Mean drained water $m^3 haá^{-1}$	%	dayáMean daily water flow (mm^{-1})
E 15/0,8	13	Tile Ø 70	10-15	3250	1800	55	3.0
	14	Tile Ø 70	10-15		1840	57	3.1
	15	Tile Ø 70	10-15		1790	55	3.0
	Average		-	-	1810	56	3.0
F 12/0,8	16	Corrugated plastic Ø 65	15-30	3250	1900	58	3.2
	17		15-30		2070	64	3.5
	18	S m o o t h plastic Ø 63 Tile Ø 70	10-15		1750	54	2.9
	Average		-	-	1907	59	3.2

From the analysis of the functional efficiency of the experimented draining variants that were analyzed during the entire period of exploitation, results the following:

- The growth of the drains depth from 0.8 to 1.0 m and keeping the spacing between the drain lines to a constant level of 12 m contributed at the decrease of the functional efficiency of the drainage system, with an average of 11 %.
- The growth of the spacing between the drain lines from 12 to 20 m and keeping of the drains depth at 0.80 m determined a decrease of the functional efficiency, with an average of 16%.
- The surface of the drain pipes joints / perforations determined the growth of the functional efficiency, with an average of 11 - 12 % (drain 17), in relation to drain 18 when the spacing between the drains was of 12 m and the medium depth was of 0.80 m (variant F).
- The association of the pipe drainage system with the land shaping in the bedding system with ridges and furrows, with an average of 23 % (variant A).

4.7 Classification and qualitative evaluation of soil

Ecologically homogenous soil-field mapping units were formed within the boundaries of the six experimental plots (Figure 1) that were prepared for subsurface drainage. The plots had the following useable areas: 0.60 ha for plots A and D, 0.45 ha for plots B and E, and 0.36 ha for plots C and F.

The classification notes for crops were determined based on the assessment of 17 eco-soil indicators. These indicators were used to evaluate the natural conditions of unimproved soil, as well as the characteristics of drained, improved, and cultivated soil. Based on the classification marks assigned to each plot, we calculated the average mark for the purposes and crops of both the control plot and the drainage plot.

Regarding quality classes, the classification scores were categorized in intervals of 20 points: I (81-100 points); II (61-80 points); III (41-60 points); IV (21-40 points); V (0-20 points).

- *Classification and qualitative evaluation of unimproved soil*

Under soil genetic conditions of the microclimate from the Baia Depression, with a *lower mean annual temperature* (7.9°C), early frost and late thawing, at which higher *mean annual rainfall* are added (806 mm), frequently stagnant water excess takes place on the field surface, at different periods of the year.

Among the important limitative factors of the production that were found in the natural conditions of albic stagnic-glossic Luvosol, used as "*hygrophilous natural meadow*", we noticed the following: stagnant moisture excess, soil acidity, soil compaction, low humus content.

Soil texture is dusty-loam at the depth 0-20 cm, and then it becomes average loam-clayey, in the subjacent layer 20-40 cm and clayey-loam deep in the soil.

Total porosity was generally correlated to the clay content (< 0.002 mm), being characterized by equal values of 52-65 % at the depth 0-20 cm and respectively, 44-47 % at the depth 20-40 cm.

Soil response is moderately acid at depth 0-40 cm.

Base saturation degree (VAh) is found within the oligomesobasic limits of 40-70 %. *The humus content* has recorded relatively high values until 194-199 t·ha-1 in the layers 0-5 and 5-18 cm. In the next two horizons EaW (18-30 cm) and EBW (36-40 cm), the humus reserve decreases once with depth until very low values.

The classification of farm field for the above-mentioned soil genetic conditions and actual use of "*hygrophilous natural meadow*" led to average classification marks of 39 points (Table 11).

Table 11. Field classification for the natural conditions of the control plot- Baia Depression.

Soil unit	Soil unit area (ha)	Agricultural use and crops	Classification note	Quality class
Unimproved	3.0	Pastures	42	III
		Hayfield	35	IV
		Average field meadows	39	IV
		Wheat	10	V
Albic stagnic- glosic		Barley	9	V
		Maize	9	V
Luvosol		Potato	10	V
		Flax	14	V
		Alfalfa	26	IV
		Average arable land	13	V

In the case of using albic stagnic-glossic Luvosol as arable land, we have evaluated the suitability of the first six crops in the wet climatic area. The average classification marks comprised between 9 points (barley and maize) and 26 points (alfalfa) led to an average of the control plot of 13 points.

The natural soil classification has shown the following defaults: water stagnation at field surface, soil stagnogleyzation, soil-ground water depth and soil acidity and compaction.

◉ *Classification and qualitative evaluation of improved soil*

By planning the subsurface drainage and applying agro-soil ameliorative works, we noticed a significant improvement of the general physical and chemical condition of the drained soil.

Soil texture was dusty-loam at the depth of 0-20 cm, while at the depth of 20-40 cm it was dusty clayey-loam, because of the mixture between genetic horizons, after soil loosening and cultivation works.

Total porosity with values between 50 and 55 %, at depths 0-40 cm, has shown a weak soil loosening after the three exploitation cycles.

Soil acidity was maintained within the limits corresponding to moderately acid to weakly acid soils, because of the long-term effect of limestone applied in 1978 and 1988.

Base saturation degree (VAh) was maintained in the first two cycles (1978 - 1986 and 1986 - 1997) between the limits of 71-90 %, moderately mesobasic. In 2010, the base saturation degree of 48-53 %, corresponding to the extremely oligomesobasic field, required limestone amendment.

Humus content that was determined after the 32-year exploitation periods, pointed out the effect of intense mineralization of humic substances.

By the application of controlling moisture excess works, the "*increase*" of classification marks of natural conditions was done.

Thus, we mention the diminution of defaults determined by classification coefficients for average annual rainfall, stagnant moisture excess and soil stagnogleyzation.

Agro-soil ameliorative works have also contributed to the "*increase*" of classification marks for the favourable correction of soil response, total porosity and organic matter content from soil. For the increased conditions of using soil for "*pastures and hayfields*" there were good average classification marks of 79-88 points, with an average of the drainage plot of 84 points that framed the improved farm field in the first quality class, with a very good fertility (table 12).

Table 12. Field classification for the increased conditions of the drainage field-Baia Depression.

Soil unit (SU)	SU area (ha)	Agricultural use and crops	Classification note	Quality class
Improved albic stagnic-glossic Luvosol	2.82	Pastures	88	I
		Hayfield	79	II
		Average field meadows	84	I
		Wheat	57	III
		Barley	50	III
		Maize	56	III
		Potato	61	II
		Flax	79	II
		Alfalfa	81	I
		Average arable land	64	II

The use of the improved soil as "arable field" framed the drained field between the average classification marks of 50 points for barley and until 81 points for alfalfa. The average of the classification marks of the drainage group for the first six crops of 64 points framed the use as "arable field" in the second quality class with a good fertility.

5. CONCISE SUMMARY

Achieving sustainable development in Romania's agricultural sector necessitates finding a harmonious equilibrium between economic advancement and safeguarding the environment. This delicate balance is contingent upon the quality of the soil and the approach taken towards land utilization.

A thorough examination of the soil in Suceava County, covering an area of 398,771 hectares, has facilitated the categorization of soil into ten distinct classes and nineteen different types. Some of the variables that restrict agricultural growth in natural soil units include excessive moisture, water erosion, acidity, flood risk, soil compaction, and low permeability soils.

A surface area of 185,316 hectares, equivalent to 21.7% of the entire area of Suceava County, experienced excessive moisture from rainfall and/or groundwater, as well as overflow from hydrographic networks, in varied forms and intensities.

The soil-units have been found to be of middling to bad quality and appropriateness for agricultural use due to the wide range of natural circumstances and inherent characteristics of the soil resources. A series of hydro-ameliorative activities were implemented over time to meet the unique needs for improving soil crust. Notable among these sorts of works are the surface systems covering roughly 55,000 hectares with ditches, as well as the pipe drainage systems spanning around 27,000 hectares. These systems are divided into 20 large and local systems.

To enhance the quality of rural areas and encourage sustainable farming practices, it is necessary to safeguard the soil by rehabilitating or expanding the current unproductive hydro-ameliorative structures and adjusting the size of agricultural plots.

The implementation of a pipe drainage system, with drain lines spaced at intervals of 12 - 15 m and a maximum spacing of 20 m, has proven to be an effective method for effectively removing surplus water.

The procedure of draining the surplus water from the soil surface and/or beneath the soil, which occurred from rainfall ranging from 40 to 160 mm over a period of 1 to 5 consecutive days, was carried out under particular conditions with a water flow rate of 1.5 to 2.5 liters per hectare per hour.

The increase in the distance between the drain lines by 20-30 meters must be linked to agricultural and soil enhancement practices, such as periodic deep loosening, mole drainage, and land shaping using ridges and furrows in the bedding system. The classification and evaluation of soil quality for various uses and crops is stored in the qualitative farm cadastre database, which is organized at the level of soil-field mapping units.

The natural conditions of albic stagnic-glossic Luvosol from the Baia Depression have been assessed. Grasslands and hayfields have an average classification score of 39 points, while arable land has a mark of 13 points. The arable land is severely limited due to excessive water.

Through the implementation of drainage systems and soil improvement techniques, we have successfully enhanced some classification indicators. Specifically, we have achieved an average score of 84 points for "improved pastures" and 64 points for "arable land".

REFERENCES

Canarache A. (1990). Fizica solurilor agricole. Ceres. Bucharest.

Canarache, A. (1991). Factors and indices regarding excessive compactness of agricultural soils. Soil & Tillage Research. Elsevier Science B.V. Publishers, 19: 145-164. Amsterdam.

Cooke, R.A., Sands, G.R., Brown, L.C. (2002). Drainage water management: a practice for reducing nitrate loads from subsurface drainage systems. *ASAE Publication*, Paper No 05. St. Joseph, Mich; 8 pp.

Dumitru, M, Ciobanu, C., Manea, Alexandrina, Gament ,Eugenia, Risnoveanu, I., Mihalache, Daniela, Tanase, Veronica, Vrinceanu, Nicoleta, Calciu, Irina, Balaceanu, Claudia, Preda, Mihaela. (2004). The evolution of main parameters of soil and agricultural lands in Romania. Soil Science National Conference. 36 A (1). pp. 39-68. Solness, Timisoara.

Florea, N., Munteanu, I. (2003). Romanian System of Soil Taxonomy, Estfalia, Bucharest, pp.30-98.

Jitareanu, G., Rusu, C., Moca, V. (2009). Evaluation and use of soil resources, environmental protection and rural development in the north-eastern part of Romania, Ion Ionescu de la Brad, Iasi, pp. 57-63.

Lukianas, A., Vaikasas, S., Malisauskas, A. (2006). Water management tasks in the summer polders of the Nemunas Lowland. *Irrigation and drainage*. 55 (2): 145-156. DOI: 10.1002/ird.230.

Man, T. E., Stoica, Fl. (2002). Amenajarea terenurilor cu exces de umiditate din vestul şi nord-vestul ţării. Sesiune Stiintifica Internatională Aniversara, Bucharest, pp. 169- 174, Bren. Bucureşti.

Mejia, M.N., Madramootoo, C.A. (1998). Improvement water quality through water table management in eastern Canada. *Journal of Irrigation and Drainage Engineering*. 124(2). pp. 116-122.

Mirsal, A.I. (2004). Soil pollution. Origin, monitoring & remediation. Springer. Berlin.

Moca, V., Canarache, A., Dumitru, Elisabeta. (1988). Modificari agrofizice ale Luvosolului albic, pseudogleic, drenat si cultivat intensiv in conditiile zonei umede din Depresiunea Baia-Moldova. Lucrarile Conferintei Nationale pentru Stiinta Solului. 26 A (1). pp. 63-76. Bucharest

Moca, V., Bofu, C., Filipov, F., Radu, O. (2001). Influenta de lunga durata a drenajului subteran si a unor lucrari ameliorative asupra solului din campul de drenaje agricole Baia-Moldova. Lucrarile Conferintei Nationale pentru Stiinta Solului. 30A (1). pp. 90-110. Bucharest

Munteanu, I., Dumitru, M., Florea, N., Canarache, A., Lacatusu, R., Vlad, V., Simota, C., Ciobanu, C., Rosu, C. (2004). Status of Soil Mapping, Monitoring, and Database Compilation in Romania at the beginning of the 21st century, *European Soil Bureau - Research Report*, 9, ISPRA Italy, pp. 251-266.

Ramoska, E., Bastiene, N., Saulys, V. (2011). Evaluation of controlled drainage efficiency in Lithuania. *Irrigation and drainage*. 60 (2). pp. 196-206

Ritzema, H.P., Kselik, R.A.L., Chanduvi, F. (1996). Drainage of irrigated lands: a manual (Irrigation water management training manual). Rome: FAO

Singh V, Frevert D, Rieker J, Leverson V, Meyer Susan, Meyer St. (2006). Hydrologic modeling inventory: cooperative research effort. Journal of Irrigation and Drainage Engineering. 132 (2): 98-103. DOI: 10.1061/(ASCE)0733-9437(2006)132:2(98)

Someren, C.L. van. (1964). De toepassing van plastieken draineerbuizen in Nederland. Cultuurtechniok, 2 Jarg. 1-3. Rotterdam.

Townend, J., Reeve, M.J., Carter, A. (2001). Water release characteristics. In: *Soil and environmental analysis: Physical methods* (2nd ed.), Smith K. A., Mullins C. E. (eds.). New York: Marcel Decker Inc.

Troeh, F.R., Thomson, L.M. (2005). Soils and soil fertility. Iowa: Blackwell Publisher.

Walker, W.R. (1989). Guidelines for designing and evaluating surface irrigation systems, *FAO Irrigation and Drainage Paper* (FAO). 45. Rome

***. (1987). Methodology of soil studies elaboration. National Institute of Research and Development in Soil Science, Agrochemistry and Environment. Bucharest, vol. I, p. 135-154,

***. (2006). World reference base for soil resources 2006. *World Soil Resources Reports.* 103. FAO. Rome.

***. (2008). Romanian Statistical Yearbook. National Institute of Statistics. Bucharest.

***. (2010). Romanian Statistical Yearbook. National Institute of Statistics. Bucharest.

INDEX

www.ingramcontent.com/pod-product-compliance
Lightning Source LLC
LaVergne TN
LVHW062225200926
848484LV00006B/1480